International Perspectives on Geographical Education

Series Editors

Now accepted by Scopus! Content available on the Scopus site in the second half of 2023.

This book series is dedicated to international research in geography education, focusing on curriculum, pedagogy and assessment related to geography education across the age range (from KG-higher education). Led by the priorities and criteria set out in the Commission's Declaration on Geography Education Research, the series plays an important role in making geography education research accessible to the global community. Publications are drawn from across the geography education research community, and also includes publications stemming from specific events supported by the Commission. Individual book author and editors are encouraged to consider how their proposals correspond to the Commission's ongoing programme of work. Research published represents immediate developments within the international geography education community. The series seeks to support the development of early career researchers in publishing high quality, high impact research accounts. This series is under the editorial supervision of the International Geography Union's Commission on Geographical Education.

Magali Hardouin • Caroline Leininger-Frézal
Editors

Curricular Changes in Geography Education

Meeting Social and Societal Challenges of Tomorrow

Springer

Editors
Magali Hardouin
Université Angers
UMR Espaces et Sociétés ESO 6590 CNRS
Rennes, France

Caroline Leininger-Frézal
Laboratory de didactique André Revuz
University Paris Cité
Athis-Mons, France

ISSN 2367-2773 ISSN 2367-2781 (electronic)
International Perspectives on Geographical Education
ISBN 978-3-031-96550-0 ISBN 978-3-031-96551-7 (eBook)
https://doi.org/10.1007/978-3-031-96551-7

This Springer imprint is published by the registered company Springer Nature Switzerland AG
The registered company address is: Gewerbestrasse 11, 6330 Cham, Switzerland

Contents

Chapter 1
Introduction Reimagining Curriculum: Geography's Response to Evolving Challenges

Magali Hardouin and Caroline Leininger-Frézal

The advent of global changes has had a profound impact on education systems worldwide, manifesting in the introduction of competency-based approaches, education for sustainable development, information and communication technologies in education (ICTE), and numerous others. A number of these changes have been driven by major international institutions, including united nations educational, scientific and cultural organization (UNESCO) and donor agencies (international monetary fund (IMF) and World Bank). Other changes are linked to societal developments. The teaching of geography has been affected by these changes, as well as by a growing disconnect between scientific geography and geography education. The extent of this rupture varies from country to country. In light of these developments, the community of geographers has reiterated the importance and legitimacy of geographical education, particularly in the aftermath of the seminal work by Young and Muller (2010), who delineated three prospective trajectories for the discipline's future in secondary education. In the first of these (Future 1), geography teaching remains encyclopedic, with knowledge taught being disconnected from the world and regarded as having an end in itself. The second scenario (Future 2) involves the prioritization of developing transversal skills within the context of geography education, albeit divorced from the epistemology of reference. The third scenario (Future 3) is characterized by teaching that is firmly grounded in a robust epistemological foundation, aligning with students' experiences and aimed at facilitating their comprehension of the world. The Geocapability theory has been

M. Hardouin (✉)
Université Angers, UMR Espaces et Sociétés ESO 6590 CNRS, Rennes, France
e-mail: magali.hardouin@univ-angers.fr

C. Leininger-Frézal
Laboratoire de didactique André Revuz, Université Paris Cité, Paris, France
e-mail: Caroline.leininger-frezal@u-paris.fr

M. Hardouin, C. Leininger-Frézal (eds.), *Curricular Changes in Geography Education*, International Perspectives on Geographical Education,
https://doi.org/10.1007/978-3-031-96551-7_1

developed to achieve this future (Solem et al., 2013). It is one of the answers to the necessity of renewing the geography curriculum, developing teaching practices, and improving student learning and teacher education. This theory has provided a foundational framework for a substantial volume of past and current research in geography education (Béneker et al., 2024; Biddulph et al., 2020; Bladh, 2020; Bustin, 2017; Fargher, 2018; He et al., 2024; KARABATAK, 2024; Mitchell, 2022; Mitchell et al., 2022; Sugiyanto et al., 2024; Uhlenwinkel et al., 2017) and serves as references for some of the chapters in this book.

The ongoing changes to the geography curriculum are significant and have gone beyond the Geocapability theory framework. These changes have impacted the intended, achieved, and implemented curriculum. An international conference was organized in 2022 in Rennes, France, under the aegis of the international geographical union (IGU) Geography Education Commission, the institut national supérieur du professorat et de l'Education (INSPE) of Brittany_UBO, and the University of Rennes 2, to examine these transformations. The event yielded three key discussion strands, which have been distilled into this book. While it does not claim to be an exhaustive record of the proceedings, it, instead, offers a concise overview of current research in geography education on curricular change. The initial area of inquiry focuses on contemporary issues and challenges in the teaching of geography, to which the initial part of the book is dedicated.

1.1 Contemporary Issues and Challenges in Geography Education

Recent curricular reforms aim to transform what students learn in order to equip them to address contemporary challenges. The range of issues they face is vast, encompassing new knowledge, content, methods, and approaches. The central challenge, however, lies in fostering a mindset in young people that diverges from that of previous generations—a task made more complex by the fact that their educators, parents, and teachers belong to those very generations.

This paradox underscores the urgency of the endeavor: encouraging students to critically examine and rethink the frameworks of thought that have contributed to the current environmental crisis, such as those centered on development, economic growth, and unregulated capitalism. Instead, new paradigms must be conceived, including concepts such as post-growth and decolonization.

The goal is no longer to have students merely reproduce what they have learned but to empower them to think creatively and independently. This is a far more ambitious and demanding objective. Consequently, curricular changes emphasize the development of advanced intellectual skills such as geographical reasoning, complex thinking, and critical analysis.

How, then, can we foster these skills in students? The four following chapters of this section are designed to serve as a foundation for teachers and trainers.

In their chapter titled "Post-Growth from the Perspectives of Young People in Germany in the Context of Education for Sustainable Development," Lara Brede, Andreas Eberth, and Christiane Meyer examine the views of 25 young Germans, aged 15–17 years, on post-growth, as expressed in focus groups. The research addresses two key questions: How do the participants position themselves in relation to selected statements critiquing the economic growth paradigm? How do the participants assess alternative approaches? The findings of the study indicate that the majority of the young people surveyed are dissatisfied with the current economic system and the economic growth paradigm. However, fewer of them advocate for the establishment of a new system. Some (n=6) believe that a change in the economic system is neither realistic nor feasible. Overall, this chapter underscores the openness of young people to a change in the economic system, as well as the role of geography education in supporting them toward this shift. The issues are somewhat different in the case of teachers.

In their chapter, Jonas Koch, Tomke van Hove, and Christiane Meyer explore the manner in which teachers and educators in general approach smart mobility. Their chapter, entitled "Critical Reflections on Smart Mobility—Raising the Awareness of Geography Educators for the Social Implications of a Digital Mobility Transition," explores three research questions: How do geography teachers and extracurricular educators comment on the current trends in (a sustainable) mobility transition?; How do they reflect on the digital aspects of mobility transition and its social implications?; To what extent does the digital mobility transition and its social implications already play a role in current formal and non-formal education and how is it implemented? The authors highlight that both teachers and educators define sustainable mobility primarily through an ecological lens, advocating for reduced car use and promoting alternatives such as public transport and cycling. Nevertheless, teachers see themselves as role models, while extracurricular educators focus on fostering critical thinking through interactive learning. Digital mobility is viewed as essential, with examples such as car-sharing and smart cities, but concerns include the exclusion of non-digital users and data security. Positive impacts of smart mobility include improved accessibility and intelligent ticketing, though its social aspects are underrepresented in education. Geography is seen as a suitable subject to explore digitalization and mobility topics. It is evident that the standpoint of the following chapter aligns with this perspective.

Xi Xiang and Michael E. Meadows are in line with this perspective and adopt a different perspective in the chapter titled Unveiling Environmental Complexity: Empowering High School Students Through Technology-Supported Simulations. They pose the question: How can a technology-supported curriculum be developed to enhance students' understanding of the complex systems associated with environmental challenges? Their study involves testing a computer-modeled fire simulation in the classroom to help students grasp the complexity of fires and the role of the environment in which they occur. The teaching scenario is grounded in an active, experiential approach. The results show that the students were able to comprehend the interactions between fire and the surrounding environment, which, in turn, enabled them to predict the fire's development. The computer simulation reduced the cognitive load on the students, allowing them to focus on the dynamics of the complex fire-environment system. Computer simulation serves as a valuable tool for understanding complexity.

In their chapter, entitled "Arguments in Geography Classes—Investigating Elementary School Students' Reasoning," Livia Reis Dantas De Souza and Sonia Maria Vanzella Castellar conducted a detailed analysis of how primary school pupils utilize geographical reasoning in order to support them. The authors operated on the basis of arguments advanced by students on the subjects of urbanization, migration, and housing conditions in the specific context of Cubatão, São Paulo, Brazil. The objective is to identify the similarities and differences between the students' reasoning and those derived from other sources. The text highlights that epistemic practices, such as argumentation, can enhance students' geographical reasoning and enrich their thinking about socio-spatial issues. By engaging in argumentation, students test their ideas, justify them with evidence, and refine their mental frameworks, improving their understanding of geographical concepts.

1.2 Tools and Learning Approaches

The section "Tools and Learning Approaches" of the book Curricular Changes in Geography Education: Meeting Social and Societal Challenges of Tomorrow focuses on innovative pedagogical tools and methodologies designed to enrich geography education. In a rapidly changing world, where global challenges and societal transformations redefine how we understand and interact with our environment, educators must continuously adapt their approaches to inspire critical thinking, foster interdisciplinary understanding, and prepare students for the complexities of the future.

This section brings together four chapters that explore novel methods for teaching geography, from leveraging cutting-edge technologies to integrating interdisciplinary and socioscientific approaches. The authors illustrate how these tools and frameworks can expand students' horizons, connect terrestrial and extraterrestrial studies, and address pressing issues such as sustainability, climate change, and civic engagement. These contributions offer practical insights and theoretical reflections on reimagining geography curricula to empower learners with the skills and perspectives needed to navigate and influence a rapidly evolving world.

Together, these works underscore the importance of innovative tools and approaches in equipping students to grapple with contemporary societal and environmental challenges, fostering informed, ethical, and engaged global citizens.

Elsa Filâtre explores in her chapter "How Can Geographic Information Systems (GIS) Increase Primary School Pupils' Awareness of Their Local" how GIS can be used to develop spatial awareness among primary school pupils in France. Drawing on research conducted in the Toulouse region, the author adopts a didactic engineering approach to guide geographical inquiries into the neighborhoods surrounding schools, using tools such as Géoportail and its "Going Back in Time" application.

The findings highlight the crucial role of GIS in geographic pedagogy. These tools enable pupils to visualize, analyze, and better understand their local environment by exploring spatial data across different scales and time periods. By integrating field trips and engaging pupils in participatory inquiries, GIS foster an active learning approach, transforming neighborhoods into dynamic spaces for investigation. This process helps pupils recognize their role as spatial actors and develop geographic citizenship based on an understanding of human practices and territorial dynamics. The author also emphasizes the challenges faced by teachers, including limited training in geography and unequal access to digital tools. Despite these obstacles, the study reveals that using GIS in education enhances pupil engagement and serves as a lever to overcome the limitations of traditional geography, often focused on abstract or distant concepts.

In continuation of this exploration of educational technologies, the next chapter broadens the analysis by focusing on the role of multisensory experiences in enriching geography education.

The chapter titled "Multi-Sensory Developments in Geography Education. A Historical Analysis of the Situation in French Educational Institutions," traces the evolution of multisensory approaches in geography education in France. Drawing on historical analysis, the authors, Jean-Pierre Chevalier and Guilhem Labinal, explore educational practices, from early geography textbooks to contemporary initiatives, highlighting techniques that engage the senses to enrich geographic learning. The chapter demonstrates that, while vision has historically dominated pedagogical practices, other senses, such as touch, hearing, and smell, have gradually been integrated into innovative approaches. These techniques include the manipulation of models, immersive school excursions, and the multisensory observation of landscapes. These methods allow students to develop a more intimate connection with their environment, transforming the spaces they study into lived experiences. The authors also emphasize the importance of field trips and hands-on workshops, which foster a holistic understanding of landscapes through varied sensory perceptions. However, the authors note that these approaches remain marginal within French curricula due to the dominance of visual tools and institutional constraints. They advocate for a reevaluation of educational practices to fully integrate multisensory dimensions into geography teaching, drawing on advancements in scientific geography and experimental pedagogical experiences. In conclusion, this chapter highlights the potential of multisensory methodologies to renew geography education by fostering a deep and sensitive connection between students and the spaces they inhabit or explore. This contribution underscores the central role of sensory experiences in building more embodied and accessible geographic knowledge.

Building on this reflection, the next chapter explores how globalized landscapes can serve as a starting point for understanding global processes and engaging students with critical issues.

In her chapter titled "Systemic Approach of Globalisation Through Worldwide Landscapes Analysis," Elisabeth Dury examines how webcams can be utilized to analyze landscapes and understand the societal challenges posed by globalization.

Drawing on research with French Lycée students, Dury explores how this innovative approach fosters knowledge and civic awareness by connecting geographic concepts with global issues. The chapter highlights the significance of landscape as a dynamic framework for interpreting global processes, particularly through the lens of coastal landscapes, which serve as archetypal global interfaces. By integrating webcam imagery, Dury demonstrates how students can analyze and compare landscapes from around the world to identify key globalization markers such as integration, territorial ranking, and environmental risks. The study emphasizes the educational potential of webcams to bridge the gap between theoretical knowledge and tangible experiences, allowing students to explore distant spaces, understand global interconnections, and critically assess the societal impacts of globalization. Despite its potential, Dury acknowledges the limitations of this approach, including the primacy of visual data over other sensory inputs and the influence of commercial biases in webcam imagery. Nevertheless, she argues that landscape analysis via webcams offers a unique means of engaging students in critical thinking and fostering civic commitment. By encouraging students to formulate and debate proposals for action, this method not only enhances their geographic literacy but also prepares them to address real-world challenges.

This critical and visual framework introduced by the analysis of globalized landscapes resonates in the final chapter, which expands the reflection to an interplanetary perspective by integrating planetary sciences into geography education.

Georgios Bampasidis and Apostolia Galani propose, in their chapter "Teaching Geo-Space Through SSI: When Earth Geography Meets Space Geography," a reconceptualization of geography curricula by integrating planetary science topics and socioscientific issues (SSI). This interdisciplinary approach connects Earth studies to planetary processes, deepening the understanding of global challenges and their societal impacts. The authors emphasize that the study of Earth's geospheres—atmosphere, hydrosphere, lithosphere, and biosphere—naturally aligns with the analysis of planetary systems. By incorporating planetary science into geography through the SSI framework, students address major issues such as climate change, resource management, and sustainability, linking terrestrial phenomena to processes observed on other celestial bodies. The chapter presents an educational framework that integrates geography and planetary exploration. A central aspect of the proposed curriculum is the SSI framework, which encourages students to engage with the ethical, social, and environmental dimensions of scientific knowledge. For example, discussions about rare earth elements highlight the economic and geopolitical challenges of resource extraction on Earth and in space. Similarly, the exploration of extraterrestrial life raises ethical questions about altering celestial environments. Pilot tests of this program with educators have demonstrated its potential to enhance interdisciplinary learning, foster critical thinking, and prepare students to tackle complex global and cosmic challenges. By connecting planetary science with geography, this approach deepens scientific understanding while promoting responsible citizenship and ethical decision-making.

The introduction of new learning tools and approaches has been combined with the introduction of new subjects to be taught.

1.3 New Subjects to Learn

Curricular changes have profoundly affected the content of geography education. First, there has been an epistemological renewal of the content and the material, aligning it more closely with scientific geography. New topics have emerged or are being addressed in different ways, such as gender, post-growth, and smart mobilities, as previously mentioned, as well as geopolitical conflicts, risks, climate change, and other related issues. This renewal is also associated with the integration of contemporary societal issues. Sustainability has played a significant role in redefining geography through the lens of human–environment interactions. These new themes are interconnected with new learning approaches. This section examines five contributions addressing the incorporation of these new teaching topics into the curriculum, including the challenges related to training.

Tal Yaar-Waisel and Sebastian Wolff-Seidel explore the challenges of training secondary school teachers in geopolitics in their chapter titled "Training Teachers to Teach Geopolitics Subjects: A Comparative Study of Israel and Germany." In this comparative study, the researchers interviewed 86 students and conducted an analysis of school textbooks and the curriculum to address three research questions: What level of basic geopolitical knowledge do geography undergraduate students in teacher training in Germany and Israel possess?; How does the curriculum of the participating students support training in the field of geopolitical knowledge? What are the differences and similarities between the two samples compared to Holms and Farber's (2002)? The study revealed significant gaps in students' geopolitical knowledge, particularly in Germany and Israel, despite students valuing the subject as vital and interesting. Teachers often avoid teaching geopolitics due to insufficient training and preparation. National standards mandate addressing these topics, but implementation in curricula is inconsistent. The study emphasizes the need for updated teaching materials, improved teacher training, and fostering critical thinking through exposure to diverse perspectives. Additionally, disparities between outdated maps and current political realities in classrooms must be addressed.

Teaching geopolitics presents a challenge for teachers as it is a relatively new subject. The introduction of such topics often requires a reorganization of the curriculum, which adds to the complexity. This issue is explored by Régis Keerle in his chapter entitled "Teaching the Concept of Territory in a Bachelor of Technology in France: Adjustments to Deal with a Bureaucratic Reform." Keerle examines the impact of a technological bachelor's curriculum reform, shaped by asymmetrical power dynamics and a competency-based approach, on the integration of geography. The findings reveal that the reform has reduced teaching time and made assessing cross-curricular skills more challenging, hindering the inclusion of geography's territorial dimension. He suggests pedagogical adjustments to better align teaching with the reform's objectives, highlighting the importance of deeper collaboration among teachers for effective implementation. This chapter demonstrates that curriculum restructuring entails a multitude of intricate modifications, extending far beyond the mere addition or removal of components. The next one is aligned also with it.

The next chapter entitled "A New Curriculum for Teaching and Learning Geography in French Elementary School. Some Reflections and Critical Notes," written by Philippe Charpentier, examines the renewal of the geography curriculum in French primary schools to make it more engaging and meaningful for students. Charpentier addresses the challenge faced by non-specialist teachers by proposing a curriculum that connects prescribed content with classroom activities. The approach centers on students' emerging questions about their daily environment rather than predetermined themes, fostering a constructivist method where students observe, question, and analyze their surroundings. The goal is to equip students with terminology, tools, and methods to develop a critical and informed worldview. Charpentier advocates for a pedagogy of slowness, focusing on a single macro-geographical issue per school period to enable in-depth and meaningful learning. This approach integrates students' personal experiences, making geography a dynamic and relevant science while supporting the development of critical thinking and preparing students to become engaged, informed adults.

This section will finally address the challenges of introducing new teaching objects for students. Ana Milanović Pešić, Marko V. Milošević, and Jovana Brankov investigate the adequacy of Geography Curricula in Serbian schools in equipping students with functional knowledge about natural disasters. The study addresses whether current educational content sufficiently prepares students to understand and respond to such hazards, with a focus on both primary and secondary education levels. Utilizing a desktop analysis of curricula from 2013 to 2018 and a survey conducted after the catastrophic May 2014 Flood, the research identifies significant shortcomings in disaster education. The findings reveal that Serbian students possess limited knowledge about natural disasters, particularly those occurring within their local context, and lack clear guidance on how to react during emergencies. The survey showed that more than half of the participants expressed a need for additional training on appropriate actions during and after disasters. This lack of preparedness underscored the necessity for curriculum reform. Consequently, the Geography Curricula were revised in 2018 to include specific outcomes aimed at addressing these gaps. The new curriculum incorporates three key learning dimensions: cognitive (understanding disaster dynamics and risks), affective (managing panic and fostering rational responses), and psychomotor (developing practical and digital skills to mitigate disaster impacts). Moreover, the revised curricula integrate tools such as GIS and disaster risk registers, enabling students to engage with real-world data and early warning systems. The study highlights that this enhanced focus on natural disasters within Geography education aligns with global best practices and contributes to strengthening societal resilience. Overall, the research emphasizes the critical role of education in shaping disaster preparedness and promoting informed, proactive behaviors among younger generations.

References

Béneker, T., Bladh, G., & Lambert, D. (2024). Exploring 'Future three' curriculum scenarios in practice: Learning from the GeoCapabilities project. *The Curriculum Journal, 35*(3), 396–411. https://doi.org/10.1002/curj.240

Biddulph, M., Bèneker, T., Mitchell, D., Hanus, M., Leininger-Frézal, C., Zwartjes, L., & Donert, K. (2020). Teaching powerful geographical knowledge – a matter of social justice: Initial findings from the GeoCapabilities 3 project. *International Research in Geographical and Environmental Education, 29*(3), 260–274. https://doi.org/10.1080/10382046.2020.1749756

Bladh, G. (2020). GeoCapabilities, Didaktical analysis and curriculum thinking – furthering the dialogue between Didaktik and curriculum. *International Research in Geographical and Environmental Education, 29*(3), 206–220. https://doi.org/10.1080/10382046.2020.1749766

Bustin, R. (2017). *An investigation into geocapability and Future 3 curriculum thinking in geography*. Doctoral thesis, UCL (University College London). https://discovery.ucl.ac.uk/id/eprint/1549680/

Fargher, M. (2018). WebGIS for geography education: Towards a geocapabilities approach. *ISPRS International Journal of Geo-Information, 7*(3), Article 3. https://doi.org/10.3390/ijgi7030111

He, Y., Tani, S., & Puustinen, M. (2024). GeoCapabilities approach to climate change education: Developing an epistemic model for geographical thinking. *Journal of Geography, 123*(2–3), 23–31. https://doi.org/10.1080/00221341.2024.2334946

KARABATAK, S. (2024). *EDUCATION & SCIENCE 2023-IV*. Efe Akademi Yayınları.

Mitchell, D. (2022). GeoCapabilities 3—Knowledge and values in education for the Anthropocene. *International Research in Geographical and Environmental Education, 31*(4), 265–281. https://doi.org/10.1080/10382046.2022.2133353

Mitchell, D., Hanus, M., Béneker, T., Biddulph, M., Leininger-Frézal, C., Zwartjes, L., & Donert, K. (2022). Enhancing teachers' expertise through curriculum leadership—Lessons from the GeoCapabilities 3 project. *Journal of Geography, 121*(5–6), 162–172. https://doi.org/10.1080/00221341.2022.2149838

Solem, M., Lambert, D., & Tani, S. (2013). Geocapabilities: Toward an international framework for researching the purposes and values of geography. *Education, 3*(3), Article 3.

Sugiyanto, Muryani, C., & Ni'matussyahara, D. (2024). Strengthening student empathy in GeoCapabilities: Digital learning innovations and pedagogical strategies for disaster mitigation. Contemporary. *Educational Technology, 16*(3), ep521. https://doi.org/10.30935/cedtech/14913

Uhlenwinkel, A., Béneker, T., Bladh, G., Tani, S., & Lambert, D. (2017). GeoCapabilities1 and curriculum leadership: Balancing the priorities of aim-based and knowledge-led curriculum thinking in schools. *International Research in Geographical and Environmental Education, 26*(4), 327–341. https://doi.org/10.1080/10382046.2016.1262603

Young, M., & Muller, J. (2010). Three Educational Scenarios for the Future : Lessons from the sociology of knowledge. *European Journal of Education, 45*(1), Article 1. https://doi.org/10.1111/j.1465-3435.2009.01413.x

Magali Hardouin is a Lecturer at the University of Western Brittany (UBO)—INSPE de Bretagne, and a researcher at the ESO-Rennes, Spaces and Societies Laboratory (UMR CNRS 6590 ESO), as well as at the CREAD at the University of Rennes 2. Her work explores the interconnections between several key dimensions of education, spatial planning, and the governance of educational systems. Her analysis focuses particularly on issues related to the trajectories and mobility of young people and students, whether local or international, and how these pathways promote skill acquisition and personal development. These dynamics are directly influenced by teaching and learning environments, educational infrastructures (construction, renovation, rehabilitation), and territorial planning, incorporating an urbanistic dimension. Furthermore, her research examines the professionalism and teaching practices, as well as the specific didactics of geography, spatial planning, and urbanism, within a context where the internationalization of training and the governance of higher education institutions play a central role.

Caroline Leininger-Frézal is a faculty member at the Université Paris Cité, affiliated with the André Revuz Didactics Laboratory. Her research encompasses the domain of geography, with a particular emphasis on socially pertinent issues and experiential teaching methodologies. She has published extensively on these topics, including the use of case studies in secondary and higher education, education for the environment and sustainable development, and the links between high school and university in history and geography. She is also known for her work on experiential education as part of the Pensée Spatiale group at the IREM in Paris.

Part I
Contemporary Issues and Challenges in Geography Education

Chapter 2
Post-Growth from the Perspective of Young People in Germany in the Context of Education for Sustainable Development

Lara Brede, Andreas Eberth, and Christiane Meyer

2.1 Introduction

Our current economic growth system is increasingly being criticized with regard to various aspects such as "incredible class and social inequalities and collapsing environmental conditions" (Harvey, 2020, 11). In 2015, the 2030 Agenda for Sustainable Development, which outlines and defines 17 Sustainable Development Goals (SDGs), was adopted by all United Nations member states (UN, 2015). Despite the criticism, SDG 8 still prioritizes economic growth as the guiding principle for our economic system. In this context, the German political economist Maja Göpel states that "the most critical aspect for turning the wheel toward fulfilling the SDGs is changing the economic paradigm" (2016, 3).

Accordingly, this chapter critically reflects on economic growth as "an integrated cultural, political, ecological, and economic process manifested as an increase in the total market value of all goods and services" (Kallis et al., 2018, 292). Subsequently, some approaches to post-growth are identified and discussed. Empowering and mobilizing youth is, among other concerns, a priority action area of Education for

L. Brede (✉)
Institute of Science Education, Geography Education Section, Leibniz University Hannover, Hanover, Germany
e-mail: brede@idn.uni-hannover.de

A. Eberth
Geography Section, Professorship for Geography with a special Focus on Education for Sustainable Development, University of Passau, Passau, Germany
e-mail: Andreas.Eberth@uni-passau.de

C. Meyer
Institute of Science Education, Leibniz University Hannover, Hanover, Germany
e-mail: meyer@idn.uni-hannover.de

M. Hardouin, C. Leininger-Frézal (eds.), *Curricular Changes in Geography Education*, International Perspectives on Geographical Education,
https://doi.org/10.1007/978-3-031-96551-7_2

Sustainable Development (ESD) and essential in the context of shaping our future (UNESCO, 2020, 32). Heilen et al. (2021) found in a quantitative study with young people, which serves as the basis for our survey, that the majority of the participants had not heard of post-growth yet, but they considered a transition of our current economic system important and expressed interest in learning more about it. Thus, we qualitatively investigated the perspectives of young people in Germany on economic growth and post-growth in our research project as this represents a research gap and can be seen as a condition for reaching the other SDGs. Some results, such as how the participants position themselves on the economic growth paradigm and alternative approaches, are illustrated and discussed in detail, enabling this research to derive implications for geography education in the context of ESD.

The research project Change instead of Growth—The SDGs and Post-Growth Economies from the Perspectives of Young People in the Context of a Societal Transformation (subproject: The SDGs and Post-Growth Economies) was funded by the Ministry of Science and Culture of Lower Saxony from 2019 to 2023.

2.2 Theoretical Framing

2.2.1 Critical Reflections on Economic Growth

Economic growth is often understood as "an increase in the goods and services produced by an economy in a given period of time, typically a year" (Victor, 2015, 109). It is measured based on the gross domestic product (GDP) (Victor, 2015). In politics, the focus increasingly shifted toward growth and thus on GDP numbers starting in the 1950s, even though GDP has been heavily criticized for obscuring unpaid work, such as care work and ecosystem services (Schmelzer, 2016). The financial and economic crisis of 2007–2008 highlighted the interconnection of various crises such as "the financial, climate, migration, hunger and biodiversity crises" (Schulz et al., 2021, 15). Subsequently, economic growth is increasingly criticized on various levels and in different disciplines as within geographical debates due to the antagonism between neoliberal capitalism and ecology with respect to the climate (Potter et al., 2012; Klein, 2014; Harvey, 2020). Neoliberal capitalism focuses on short-term profitability and maximum benefits from unregulated resource exploitation and wasteful production, contradicting climate protection advocates' long-term ecological focus on sustainability and environmental preservation. Accordingly, sustainable development is not possible within the current economic system (Potter et al., 2012).

Furthermore, economic growth negatively impacts social equality as well as the environment and climate change. Indeed, as some critics have noted, "capitalist economies grow by using more resources and more people, more intensively. Accelerating this is unlikely to spare resources" (Kallis et al., 2018, 296). This incongruousness leads researchers like Tim Jackson to question the growth paradigm, demanding an understanding of prosperity without growth and advocating a system change (2009). Accordingly, finding and developing alternative business models and practices are urgently needed endeavors (Paech, 2012). However, SDG 8 does not promote a new way of thinking about the economy; instead, it embeds the current economic paradigm into a more sustainable frame without suggesting ideas for appropriate implementation (Eberth et al., 2023).

2.2.2 *Designs for a Post-Growth Era*

Post-growth encompasses various alternative approaches characterized by a detachment from the current economic growth system. It is not to be equated with contraction or recession, but rather breaks with the notion that increased efficiency and technological advances allow for endless growth in production and consumption (Schulz et al., 2021). It is about "examining the long-term meaningfulness of certain developments, and, if necessary, looking for possible alternatives within free social conditions. Meaningfulness refers here not only to the environment but also to individual and social needs, i.e. focus [on] the common good rather than individual profitability" (ibid., 20).

Post-growth, however, does not wholly abandon the growth discourse. Accordingly, and alongside the previous term, "degrowth" represents the concept of an alternative economic pathway—yet, unlike "post-growth," it is uncoupled from the discourse of growth. Degrowth is defined as "a process of political and social transformation that reduces a society's throughput while improving the quality of life" (Kallis et al., 2018, 292). Despite the lack of clarity in scientific discussions regarding the relationship between post-growth and degrowth, we consider post-growth, and especially post-growth economies (in the plural), as an economic facet of the degrowth discourse (ibid.).

However, post-growth is criticized for various concerns regarding its feasibility. These include fears of rising unemployment due to the lack of economic growth, the absence of alternatives to GDP, and the challenge of securing pensions, particularly in the context of demographic change (Strunz & Schindler, 2018). Although some post-growth approaches may seem utopian, they offer important starting points for a transition toward a society living without growth (Kallis et al., 2018). Thus, they enable the emergence and development of opportunities for action.

Actions in post-growth economies can occur on different levels (Fig. 2.1). The system and target knowledge is transmitted via several new models and theories, which serve as a basis for a transition, such as the post-growth economy (in the singular) by Paech (2012). Regarding practices of action that transfer knowledge of transformation, so-called change agents are crucial (Eberth et al., 2023). These "pioneers of change" (Eberth et al., 2023, 159) are individuals, groups, initiatives, or companies that challenge business as usual and implement alternative practices to disrupt existing mindsets and behaviors "as well as providing others who think as they do (followers, early adopters) with a constant motivation for self-sustaining change" (WBGU, 2011, 243).

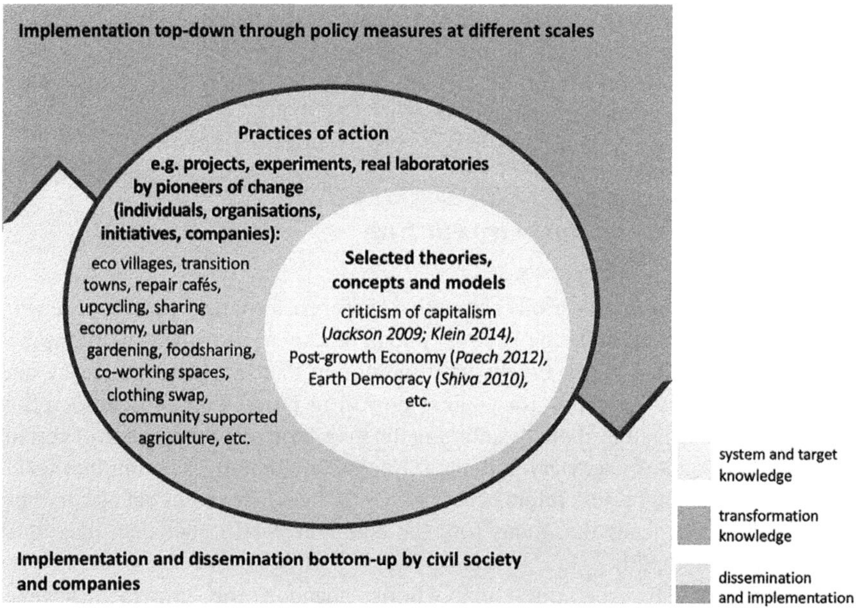

Fig. 2.1 Selected approaches in the field of post-growth economies (adapted from Eberth et al., 2023, 159)

The initiatives mentioned above are pivotal in the context of a socio-ecological transformation. Moreover, young people's involvement is also central in discussions about shaping our future (UNESCO, 2020).

2.2.3 Education for Sustainable Development and Transformative Literacy

Building on young people's geographies (Lambert, 2009), the examination of the young people's perspective is essential and also linked to ESD, referring to young people as active participants who must be encouraged "to explore values which could provide an alternative to consumer societies, such as sufficiency, fairness and solidarity" (UNESCO, 2020, 58). ESD should more clearly address the unsustainable production methods of our current economic system, thereby empowering and encouraging people to become politically active, for example, advocating for environmentally friendly regulations for businesses (ibid. 2020).

In this context and in reference to the priority action area "Empowering and mobilizing youth" (UNESCO, 2020, 32), ESD learning should happen in and beyond school—in students' everyday lives—to promote skills such as social learning. Therefore, a new understanding of the role and function of schools is necessary (UNESCO, 2020).

To involve young people's perspectives in shaping our future, transformative learning can be seen as a tool to help young people become change agents. Transformative learning is understood as a profound shift that "involves our understanding of ourselves and our self-location; our relationships with other humans and with the natural world" (O'Sullivan, 2012, 164). Crucially, O'Sullivan (2012) elaborates that this form of learning involves among other things "our visions of alternative approaches to living, and our sense of possibilities for social justice, peace and personal joy" (164).

Transformative learning requires transformative education. Organizations such as United Nations Educational, Scientific and Cultural Organization (UNESCO) have acknowledged this; for example, the ESD for 2030 roadmap refers to a "great individual and societal transformation required to address the urgent sustainability challenges" (UNESCO, 2020, 3). In each aspect of ESD, transformative learning requires an "action-oriented, transformative pedagogy [that] supports self-directed learning, participation and collaboration, problem-orientation, interdisciplinarity and transdisciplinarity, and the linking of formal and informal learning" (UNESCO, 2017, 7). The transformational process of education views the interaction between the individual and their environment as the individual's response to new challenges, thus involving transformative literacy (Koller, 2020).

Transformative literacy serves as an orientation for transformative education (Meyer & Höbermann, 2021) and consists of system knowledge (i.e., problem analysis), target knowledge (i.e., visions for the future), and transformation knowledge (i.e., understanding the conditions needed for the realization of visions) (see Fig. 2.1). In this context, the learners are "[b]uilding capacity to think critically about [and beyond] what experts say and to test sustainable development ideas" (Vare & Scott, 2007, 194). Thus, transformative literacy matches the principles of ESD (Singer-Brodowski & Schneidewind, 2014; Meyer & Höbermann, 2021). The concept of transformative literacy serves as the basis for our study's research design, presented in the following section.

2.3 Research Design: Questions, Methods, and Sample

To examine young people's ideas, attitudes, and evaluations on post-growth, we will focus on two primary research questions from our research project:

- How do the participants position themselves on selected statements criticizing the economic growth paradigm?
- How do the participants comment on a system change toward post-growth?

The research design consisted of three meetings conducted in 2020 and 2021 outside of the school context and during the COVID-19 pandemic. These meetings were carried out remotely via the online platform Webex. The first and final meetings were conducted as focus groups (Bedford & Burgess, 2001) (see Fig. 2.2),

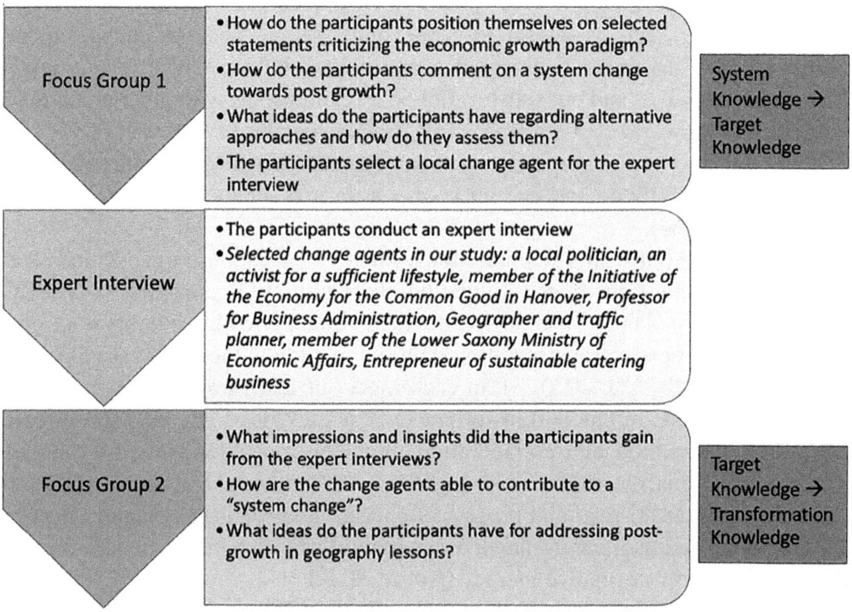

Fig. 2.2 Three-part structure of the survey (own illustration)

"a research technique that collects data through group interaction on a topic determined by the researcher" (Morgan, 1996, 130).

Through the outlined research questions, this chapter focuses on content gleaned from the first focus group meeting, which pertained to system and target knowledge. The study design assumed that the participants would not have been in contact with post-growth topics before (Heilen et al., 2021). Therefore, this set of focus group discussions was structured with the help of a prepared guideline and specific stimuli. Initially, the participants were asked to comment on a picture of a Fridays for Future demonstration and the slogan "System Change, not Climate Change," two items linked to the everyday experiences of young people. Afterward, they watched an extract of a speech by Greta Thunberg at the COP 24 climate conference in Katowice, analyzed diagrams on the Anthropocene and The Great Acceleration by Steffen et al. (2015), and stated their positions. This process was chosen to enable a direct contrast between an emotion-laden speech and a scientific source. Subsequently, they chose to read and offer their opinions on one of four short texts: (1) Niko Paech (excerpt on the concept of a post-growth economy), (2) Tim Jackson (excerpt on prosperity without growth), (3) Naomi Klein (excerpt on the Green New Deal), and (4) Vandana Shiva (perspective from an actor of the so-called Global South). Based on a definition of post-growth economies, they discussed the possible characteristics of a system change (Eberth et al., 2022). At the end of the first focus group meeting, the participants identified a local change agent with whom they wanted to conduct an expert interview.

Table 2.1 Overview of the study sample

	Number of participants	Gender distribution
Group 1	4	4 male, 0 female, 0 other
Group 2	3	0 male, 3 female, 0 other
Group 3	4	2 male, 2 female, 0 other
Group 4	3	3 male, 0 female, 0 other
Group 5	4	4 male, 0 female, 0 other
Group 6	4	4 male, 0 female, 0 other
Group 7	3	1 male, 2 female, 0 other
Total	25	18 male, 7 female, 0 other

Between 2 and 3 weeks after the first online focus group discussion, each focus group conducted a digital expert interview with the selected change agent with regard to their contributions to a sustainable transformation. In this case, the experts—change agents—are defined as people who command a particular knowledge on a topic through experience, insight, or status (Meuser & Nagel, 2009) (Table 2.1).

Approximately 1 week after the expert interview, the final meetings and second focus group discussions occurred online. As this discussion focused on target knowledge but mainly on transformation knowledge, these meetings were more open and less structured than the first focus group.

Each of the focus groups' three meetings were recorded and transcribed. Afterward, the data were analyzed using the Qualitative Content Analysis approach by Kuckartz (2019). For this, deductive main categories derived from central theoretical findings have been formed. Those have been complemented through inductive subcategories. The software MAXQDA was employed for the coding and analysis.

To answer the above questions, we recruited 25 young people between 15 and 17 years old from the Hanover region in Northern Germany to participate in the survey. The age group was chosen as the core curriculum of the upper grades is compatible with ESD-related topics and the SDGs (Niedersächsisches Kultusministerium, 2017). As gender did not have an effect on the argumentation, the gender relation of the participants must not be seen as problematic (see Table 2.1).

2.4 Selected Results of the Empirical Study

This chapter provides insights into the content of the first set of focus groups, focusing on the position of the participants toward the economic growth paradigm and alternative approaches. Crucially, the participants' statements must be seen in connection to the outlined stimuli. To examine the first research question, we guided the participants to position themselves on the four outlined texts. Following the discussion of the texts, the answers in the context of the second research question must also be considered in light of the definition of post-growth drawing on the explanations in Chap. 3.

2.4.1 How Do the Participants Position Themselves on Selected Statements Criticizing the Economic Growth Paradigm?

In commenting on the statements, most of the young people expressed dissatisfaction with the current economic growth paradigm. They mentioned negative impacts on the environment, as their dissatisfaction was mainly justified by capitalism's ignorance of the effects on nature and the climate.

"Capitalism ensures that we always want more and that the economy grows even though that does not work with our earth and our resources" (Danil_I_14).

Aside from the growth paradigm's impacts on the environment and the climate, it is striking that the participants frequently mentioned their dissatisfaction with two products of this paradigm: social division and privileged groups' luxurious lives. Indeed, as one participant noted,

"Growth is the main cause for growing inequality. In my opinion, it can be transferred to any kind of inequality: in society, between rich and poor, [and] between available resources and those that we consume" (Sanja_I_42).

Social division is closely intertwined with global injustice. The participants stated that they see the current economic growth system as a fundamental reason for exploitation and the unbalanced distribution of prosperity.

"Capitalism induces [us] to exploit people in other regions that work under bad working conditions and without respecting any environmental guidelines, which promotes climate change" (Emilia_I_39).

The dissatisfaction with the current use of resources was frequently mentioned aspect by this study's participant group—young people—when criticizing the economic growth system.

"Special attention is always given to technological, medical and industrial progress. But the crucial point is the nature. … It should be about the resources that are used; they will not be there forever. They will end at some point, [so] we have to do something now" (Lia_I_159).

In summary, the majority of the participants agreed with the criticism in the statements and supported it with the following arguments: consequences for the environment due to resource exploitation, climate change, social division between rich and poor, and global injustice.

Four participants did not agree with the criticism in the statements. They argued that growth is vital for a functioning economic system. In this case, their argumentation was often based on the assumption that growth automatically accompanies equality and prosperity. Furthermore, they focused on countries of the so-called Global North—and often, only on Germany. In the context of global justice or injustice, consequences or developments in other regions were not acknowledged or considered.

"I think that our current economic system as it is now has led to equality through economic upswing and welfare. Capitalism implies prosperity. I do not think that this has led to inequality within our society" (Markus_I_34).

"I think [the economy with] growth as its foundation is also based on efficiency. That makes it competitive and better than other economies, which leads to growth again" (Lars_I_30).

2.4.2 How Do the Participants Comment on a System Change Toward Post-Growth?

Following the definition of post-growth, many participants explicitly demanded a change in the current economic system. Those participants demanding a change in the economic system suggested different possibilities for action. The most frequently mentioned aspects in this context were the reduction of consumption, waste separation, and prevention, as well as saving resources. Furthermore, it is striking that the participants who demanded a system change on the economic as well as societal, individual, and political levels showed a greater dissatisfaction with older generations than the participants who did not demand a system change. Regarding this aspect, intergenerational justice and the feeling of not being taken seriously by older people played an essential role in the participants' views. In the context of a system change, they expressed that this makes them feel demotivated and prevents them from taking action.

"When young people deal with topics like sustainability or Fridays for Future, I feel like they are being mocked by older people. ... They say that pupils are only part of Fridays for Future so that they can skip school. I also think that many pupils and young people are not really interested in this topic because they do not take it seriously as older people and authority figures depreciate the engagement for our future" (Samira_I_169).

In many cases, the participants' argumentation for a different economic system was combined with the claim of political interference and a fundamental change. In turn, the demand for a fundamental change was closely intertwined with the need for an individual change.

"Everyone has to do one's bit and reduce [their] own consumption. [C]onsumption and, therefore, economic growth cause more and more problems, which makes it hopeless. ... But also politics, laws, legislation... There must be changed a lot as well" (Sanja_I_49).

In the context of a system change toward a post-growth economy and society, the importance of role models and self-efficacy were vital aspects for the participants. It was frequently mentioned that role models, for example, local change agents, are pivotal in increasing one's motivation to become self-effective.

"I think that people have to get motivated and have to be shown … that it is a movement. Then more and more people will join because it gives them a sense of belonging" (Sanja_I_52).

In summary, most participants at some point mentioned aspects that plead for a system change and alternative approaches either on a societal and fundamental, individual, economic, or political level.

Some participants critically stated that they did not consider a new economic system, respectively, a system change as conceivable and feasible. The participants justified this position by saying that growth is essential for a functioning economic system, that it has brought prosperity and technological progress, and that Germany cannot achieve it independently without being economically disadvantaged.

"If a country like Germany will economize according to post-growth, but another country like China does not abide [by] the rules, they will benefit from their old economic system as they lose Germany as their [economic] competitor" (Brian_I_58).

2.5 Discussion and Critical Reflection

This study found that most participants expressed dissatisfaction with the current economic growth system due to the harmfulness to the environment, the use of resources, climate change, social division, the luxurious life of some people, and global injustice. In their argumentation, they agreed with the criticism of the growth paradigm (Kallis et al., 2018). In this context, it must be emphasized that the participants mentioned the negative impacts of economic growth on an ecological level while considering social aspects. Some participants saw the negative impacts of economic growth on social equality in their region while broadening their view of the global effects of economic growth. Conversely, the participants supporting the growth paradigm overlooked this global perspective. Furthermore, they argued that prosperity naturally accompanies growth and conveys social equality. Contrary to these views, research shows that the connection between economic growth and prosperity is much weaker than is often assumed (Demailly et al., 2013). The participants advocated for a system change toward post-growth, identified general courses of action, and perceived responsibility both at the political and individual levels. Nevertheless, the argumentation remained rather superficial, as the participants first had to be made aware of the underlying problem, leaving little time for discussing alternative approaches and a system change. The absence of specific action proposals in this context highlights the significance of engaging with change agents.

It can be critically noted that, within the scope of this study, the young people were only provided with a selective view of the topic. Therefore, selected aspects of post-growth should have been explored in more depth, for instance, through the presentation and discussion of concrete approaches such as Paech's post-growth economy and other practices of action (Fig. 2.1).

2.6 Conclusion

The theoretical framework of this work has shown that this topic is crucial within the context of sustainable development. Even though the scope of the study is demographically and geographically limited, some implications for geography education can be derived if considering the criticisms about economic growth, alternatives to the predominant growth paradigm, and the results of our research. Although most of the participants were critical of the current economic growth paradigm, the opinions and attitudes of the young people were widely divergent.

First, it must be stated that the participants who supported the growth paradigm did not broaden their view of global consequences, which are central criticisms as outlined in this chapter. Therefore, further clarification in this global context is essential. Second, role models are seen as crucial for self-effectiveness by the participants. Consequently, their visibility on a regional level should be increased. This is linked to the central action fields of ESD (UNESCO, 2020) and is of great relevance to motivate young people, who levy the critique that older generations do not take them seriously, to become actively involved. A differentiated engagement with this topic within school education is necessary, one that focuses on concrete local initiatives but also integrates a global perspective, such as the exploitation during the colonial era. Drawing on ESD, the topic offers opportunities for inter- and transdisciplinary teaching (UNESCO, 2020). In conclusion, it can be stated that research in this context is still in its early stages. This study has provided an initial insight; however, further research in this field is necessary.

References

Bedford, T., & Burgess, J. (2001). The focus-group experience. In M. Limb & C. Dwyer (Eds.), *Qualitative methodologies for geographers. Issues and debates* (pp. 121–135). Arnold.

Demailly, D., Chancel, L., Waisman, H., & Guivarch, C. (2013). *A post-growth society for the 21st century. Does prosperity have to wait for the return of economic growth?* Studies n°08/13. Retrieved October 2, 2023, from https://www.greens-efa.eu/legacy/fileadmin/dam/Documents/Events/2013_11_14_A_post_growth_society/0003-Study0813-EN-WEB.pdf.

Eberth, A., Meyer, C., & Heilen, L. (2023). "Economics for Future" from different perspectives: Critical reflections on SDG 8 with a special focus on economic growth and some suggestions for alternative pathways. In M. L. De Lázaro Torres & R. De Miguel González (Eds.), *Sustainable development goals in Europe: A geographical approach* (pp. 153–167). Springer Nature Switzerland AG.

Eberth, A., Meyer, C., & Heilen, L. (2022). Fokusgruppen zu Postwachstumsökonomien im Kontext transformativer Bildung. In A. Eberth, A. Goller, J. Günther, M. Hanke, V. Holz, A. Krug, K. Roncévic, & M. Singer-Brodowski (Eds.), *Bildung für nachhaltige Entwicklung – Aktuelle Impulse zu Digitalisierung, Inklusion und Klimaschutz. (Ökologie und Erziehungswissenschaft, Schriftenreihe der BNE-Kommission in der DGfE)* (pp. 144–164). Barbara Budrich.

Göpel, M. (2016). *The great mindshift. How a new economic paradigm and sustainability transformations go hand in hand.* Springer.

Harvey, D. (2020). *Anti-capitalist chronicals*. Pluto Press.

Heilen, L., Eberth, A., & Meyer, C. (2021). Young people's geographies in the times of Covid-19: System threat as a chance for system change? In C. Travis & V. Valentino (Eds.), *Narratives in the anthropocene era* (pp. 302–320). Il Sileno.

Jackson, T. (2009). *Prosperity without growth. Economics for a finite planet*. Earthscan.

Kallis, G., Kostakis, V., Lange, S., Muraca, B., Paulson, S., & Schmelzer, M. (2018). Research on degrowth. *The Annual Review of Environment and Resources, 43*, 291–316. https://doi.org/10.1146/annurev-environ-102017-025941

Klein, N. (2014). *This changes everything: Capitalism vs the climate*. Simon & Schuster.

Koller, H.-C. (2020). Problems and perspectives of a theory of transformational processes of bildung. *Educational Theory. Symposium: Transformative Education: Philosophical, Psychological, and Pedagogical Dimensions, 70*(5), 633–651. https://doi.org/10.1111/edth.12448

Kuckartz, U. (2019). Qualitative text analysis: A systematic approach. In G. Kaiser & N. Presmeg (Eds.), *Compendium for early career researchers in mathematics education* (pp. 181–197). Springer.

Lambert, D. (2009). Introduction – part 1: What is living geography? In D. Mitchell (Ed.), *Living geography. Exciting futures for teachers and students* (pp. 1–7). Chris Kington Publishing.

Meuser, M., & Nagel, U. (2009). The expert interview and changes in knowledge production. In A. Bogner, B. Littig, & W. Menz (Eds.), *Interviewing experts* (pp. 17–42). Palgrave Macmillan.

Meyer, C., & Höbermann, C. (2021). From fast fashion to slow fashion – Raising awareness of young people for sustainable production and consumption. In M. Fargher, D. Mitchell, & E. Till (Eds.), *Recontextualising geography in education* (pp. 167–183). Springer Nature Switzerland.

Morgan, D. L. (1996). Focus groups. *Annual Review of Sociology, 22*, 129–1523. https://doi.org/10.1146/annurev.soc.22.1.129

Niedersächsisches Kultusministerium (Ed.). (2017). *Kerncurriculum für das Gymnasium – gymnasiale Oberstufe, die Gesamtschule – gymnasiale Oberstufe, das Abendgymnasium, das Kolleg*. Erdkunde. Unidruck.

O'Sullivan, E. (2012). Deep transformation: Forging a planetary worldview. In E. W. Taylor & P. Cranton (Eds.), *The handbook of transformative learning. Theory, research, and practice* (pp. 162–177). Jossey-Bass.

Paech, N. (2012). *Liberation from excess. The road to a post-growth economy*. oekom verlag.

Potter, R., Conway, D., Evans, R., & Lloyd-Evans, S. (2012). *Key concepts in development geography*. SAGE.

Schmelzer, M. (2016). *The hegemony of growth: The OECD and the making of the economic growth paradigm*. Cambridge University Press.

Schulz, C., Lange, B., Huelz, M., & Schmid, B. (2021). Post-growth geographies. Conceptual and thematic cornerstones of this book. In B. Lange, M. Huelz, B. Schmid, & C. Schulz (Eds.), *Post-growth geographies spatial relations of diverse and alternative economies* (pp. 15–36). Transcript.

Singer-Brodowski, M., & Schneidewind, U. (2014). Transformative literacy. Gesellschaftliche Veränderungsprozesse verstehen und gestalten. In FORUM Umweltbildung im Umweltdachverband (Ed.), *Krisen- und Transformationsszenarios: Frühkindpädagogik, Resilienz & Weltaktionsprogramm* (pp. 131–140). Forum Umweltbildung.

Steffen, W., Broadgate, W., Deutsch, L., Gaffney, O., & Ludwig, C. (2015). The trajectory of the Anthropocene: The great acceleration. *The Anthropocene Review, 2*, 1–18. https://doi.org/10.1177/2053019614564785

Strunz, S., & Schindler, H. (2018). Identifying barriers toward a post-growth economy – A political economy view. *Ecological Economics, 153*, 68–77.

UNESCO. (2017). *Education for sustainable development goals: Learning objectives*. https://www.unesco.de/sites/default/files/2018-08/unesco_education_for_sustainable_development_goals.pdf.

UNESCO. (2020). *Education for sustainable development. A roadmap*. https://unesdoc.unesco.org/ark:/48223/pf0000374802.locale=en.

UN: United Nations. (2015). *Transforming our world: The 2030 agenda for sustainable development*. Resolution adopted by the general assembly on 25 Sept 2015 (A/RES/70/1). https://sdgs.un.org/2030agenda.
Vare, P., & Scott, W. (2007). Learning for a change. *Journal of Education for Sustainable Development, 1*(2), 191–198. https://doi.org/10.1177/097340820700100209
Victor, P. (2015). Growth. In G. D'Alisa, F. Demaria, & G. Kallis (Eds.), *Degrowth. A vocabulary for a new era* (pp. 109–112). Routledge.
WBGU: German Advisory Council on Global Change. (2011). *World in transition. A social contract for sustainability*. https://www.wbgu.de/fileadmin/user_upload/wbgu/publikationen/hauptgutachten/hg2011/pdf/wbgu_jg2011_kurz_en.pdf.

Lara Brede has been a research assistant and PhD student since 2022 in the Geography Education section of the Institute of Science Education at Leibniz University Hannover. From 2016 to 2022, she pursued a training program to become a teacher in secondary schools (Gymnasien). Her current research focuses on Education for Sustainable Development, Post-Growth, and Transformative Learning.

Andreas Eberth is a Professor Dr. at the University of Passau, holding the Chair of Geography with a focus on Education for Sustainable Development. His research interests include Education for Sustainable Development, Transformative Education, Visual Geographies, Postcolonial Perspectives in Geography Education, with a particular focus on East Africa.University of Passau, Professorship for Geography with a Focus on Education for Sustainable Development; Research Interests: Education for Sustainable Development, Transformative Education, Visual Geographies, Postcolonial Perspectives in Geography Education, East Africa.

Christiane Meyer has held the Professorship for Geography Education at the Institute of Science Education at Leibniz University Hannover since 2008. From 1990 to 1998, she pursued studies and preparatory service for teaching at grammar schools (Gymnasien), followed by a year as a teacher from 1998 to 1999. Between 1999 and 2008, she worked as a research assistant and completed her PhD at the University of Trier. Her current research interests include Education for Sustainable Development, Transformative Learning, values education, and cultural awareness.

Chapter 3
Critical Reflections on Smart Mobility. Raising the Awareness of Geography Educators for the Social Implications of a Digital Mobility Transition

Jonas Koch, Tomke van Hove, and Christiane Meyer

3.1 Introduction

Undoubtedly, transformative change in the mobility sector is imperative to achieve the set climate goals. There is currently great hope that smart mobility solutions can encourage the use of more sustainable transport options. The concept of smart mobility is based on the idea of digitally connecting different mobility services, in order to make multimodal transport behavior more attractive and to reduce car dependency in favor of shared mobility (Groth, 2019). However, this trend poses a significant risk of leaving less tech-savvy groups behind by potentially reducing their ability to use mobility services. Education is a key influencer of norms and value systems, and formal and non-formal education systems play a central role in paving the way for more sustainable mobility in the future. Teaching about smart city, smart mobility, and their social implications should, therefore, be an integral part of Education for Sustainable Development (ESD), especially in geography education. This is only possible if teachers and extracurricular educators themselves are aware of the current developments, including their advantages and disadvantages, and have high-quality learning materials at their disposal. As Dorsch (2018) notes, this has not been the case so far, as the topic has not yet been didactically adapted. Therefore, we aim to begin to bridge that gap within our research project: "Societal Transformation in a Digital Society – Raising Awareness for the Participation in a

J. Koch (✉) · C. Meyer
Institute of Science Education, Leibniz University Hannover, Hannover, Germany
e-mail: jonas.koch@zuv.uni-hannover.de; meyer@idn.uni-hannover.de

T. van Hove
Lower Saxony Water Management, Coastal and Nature Protection Agency, Hannover, Germany
e-mail: tomke.vanhove@nlwkn.niedersachsen.de

M. Hardouin, C. Leininger-Frézal (eds.), *Curricular Changes in Geography Education*, International Perspectives on Geographical Education,
https://doi.org/10.1007/978-3-031-96551-7_3

Sustainable and Inclusive Mobility Transition." The research project consists of four parts, each focusing on a different perspective in the urban regions of Hannover and Hamburg (Germany): Teachers and extracurricular educators, secondary school students (15–16 years old), local decision-makers, and public transport users.

In this chapter, we present our findings on the perspectives of school teachers and extracurricular educators in Hamburg and Hannover (Germany) on smart mobility and its social aspects. We begin this chapter by presenting the background of current trends in mobility transition, followed by contextualizing the role of ESD within this topic. We continue with a description of the study design, which leads to the presentation and discussion of the results. The chapter concludes with a summary of the main findings and an outlook.

3.2 The Necessity of an Inclusive Mobility Transition

The Sixth Assessment Report of the Intergovernmental Panel on Climate Change makes it clear that: "Meeting climate mitigation goals would require transformative changes in the transport sector" (Jaramillo et al., 2022: 1051). In 2019, the transport sector accounted for 23% of all energy-related emissions (8.9 Gt CO2eq), of which 70% were released by road vehicles (IEA, 2023b). Between 1990 and 2022, global transport emissions grew by an average of 1.7% per year (IEA, 2023a). It is becoming clear that the transport sector is not on track to meet the vision of net-zero carbon emissions by 2050 set out in the Paris Climate Agreement (Axsen et al., 2020).

Most of the proposed measures to reduce CO2 emissions in the transport sector can be categorized under the Avoid-Shift-Improve (ASI) framework (see Fig. 3.1). While Avoid strategies aim to reduce the total number of vehicle trips (e.g., by working from home), Shift strategies seek to encourage the use of less harmful transport options (e.g., public transport). Improve strategies target advances in vehicle technology (e.g., electric cars) (Scherf & Wolter, 2016). Although all three components of the ASI framework play an important role, Avoid strategies have the highest potential to lower emissions (Jaramillo et al., 2022).

Mobility is an essential part of everyday life and a critical prerequisite for meeting basic human needs (e.g., education or participation in society) (Preston & Rajé, 2007). The transport-related social exclusion approach highlights that some social groups are excluded from essential activities due to a lack of access to mobility options, which, in turn, leads to social exclusion and reduced quality of life (Kenyon et al., 2002; SEU, 2003). In this context, John Urry and Mimi Sheller have coined the term mobility justice, which emphasizes the idea that unequal access to mobility opportunities perpetuates and even reinforces existing social inequalities (Sheller, 2020). One of the key questions in this regard is how different social groups are affected by the current mobility transition, and how meaningful change is possible to create more equitable and inclusive transport options (Lucas, 2019).

Currently, the global transformation of mobility is being driven by digitalization (Parkhurst & Seedhouse, 2019). This is based on the hope that smart mobility

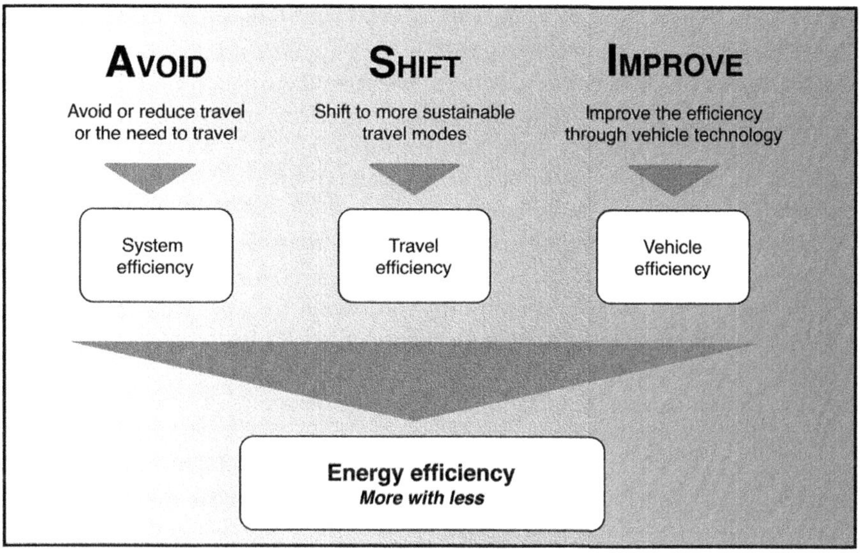

Fig. 3.1 The ASI framework for sustainable mobility (own representation with reference to Scherf & Wolter, 2016)

solutions can support a shift away from private cars to more sustainable transport options by making multimodal mobility behavior more convenient (Smith et al., 2018). These promises need to be critically evaluated (Kamargianni et al., 2016). If digital mobility solutions are successful and become the dominant or even the only access point to (public) transport networks, there is a real risk of "technological gentrification" (Pangbourne et al., 2020: 43). As the digital divide runs along many factors such as gender, age, income, race, and location, the reliance on digital solutions and the introduction of algorithmic decision-making pose real risks and could contribute to mobility injustice (van Dijk, 2020). It is, therefore, important to broaden the focus of research from the distribution of transport to recognizing the needs of different user groups (Verlinghieri & Schwanen, 2020). This topic has been taken up by mobility research the German-speaking countries in recent years (Groth, 2019).

3.3 Enabling Change Through Geography Education in the Context of ESD

In 2015, the United Nations adopted the 2030 Agenda for Sustainable Development and its 17 Sustainable Development Goals (SDGs). ESD has been identified as a key instrument for achieving these SDGs (UNESCO, 2017). Therefore, in 2019, the United Nations Educational, Scientific and Cultural Organization (UNESCO) introduced a new global framework on ESD: "Education for Sustainable Development: Towards achieving the SDGs (ESD for 2030)" (UNESCO, 2020: 55). Achieving the SDGs depends to a large extent on the collective efforts of global societies. Following the theory of social tipping elements by Otto et al. (2020), which is based on the idea that interventions in social subsystems can lead to changes at the global level, the education system (formal and non-formal) plays a crucial role in social transformation processes. In this context, the education system is the basis for understanding cause and effect, as well as possible mitigation measures.

Geography education is particularly important for the implementation of ESD. Geography as a discipline has the advantage of a holistic approach, including natural and social sciences and their interactions, with a particular focus on places and spaces. It is extremely versatile in integrating the topics of climate change and sustainable development into the school curriculum and offers strong anchor points for extracurricular learning in the field of ESD (Meadows, 2020). While national geography curricula vary, the Commission on Geographical Education of the International Geographical Union clearly states that these goals apply to geography education worldwide (IGU-CGE, 2016).

Teachers play a key role in the implementation of ESD as their perspective is decisive for learner-centered approaches and action-oriented learning. Their knowledge, skills, attitudes, and values are important for "Priority Action Area II: Transforming learning environments" and "Priority Action Area III: Building capacities of educators" (UNESCO, 2020). Several studies have shown the tremendous impact of educators, providing empirical evidence that certain attitudes of teachers and their credibility promote influence within the framework of educational processes (Hattie & Zierer, 2019). Respecting this framework, teachers can be described as change agents for a societal transformation (UNESCO, 2017). This also applies to extracurricular educators and non-formal learning places, as they provide important additional knowledge from experts outside of the school context. Non-formal learning places lead to greater student engagement and can act as a stimulus. Therefore, a strong connection between formal and non-formal education is paramount for ESD (UNESCO, 2020). In the state of Lower Saxony, for example, more than 60 recognized extracurricular learning sites for ESD can be identified, including geoparks and environmental education centers (LSE, 2023).

In the context of our study, it is important to acknowledge that young people are active users of (public) transport systems and early adopters of mobility innovations, with specific requirements for their daily mobility (Casadó et al., 2020; UNECE, 2020). By equipping them with knowledge on mobility transition, smart mobility, and mobility justice and by promoting self-determined engagement, young people can significantly support the transformation toward more inclusive, safe, resilient, and sustainable mobility as a key aspect of SDG 11 on sustainable and inclusive cities and communities (Kanning & Meyer, 2022; Meyer & Haubner, 2022).

3.4 Study Design and Methods

We conducted interviews with geography teachers and extracurricular educators in the cities of Hannover and Hamburg in northern Germany. Both cities have a range of upcoming digital mobility services available, making them excellent case studies. Due to the selected study areas, we focused on urban mobility. We recognize that the results and answers could have been significantly different in the context of rural regions.

With respect to the overall aim of our study, we focused on three research questions:

1. How do geography teachers and extracurricular educators comment on the current trends in (a sustainable) mobility transition?
2. How do they reflect on the digital aspects of mobility transition and its social implications?
3. To what extent does smart mobility and its social implications play a role in current formal and non-formal education and how is it implemented?

We conducted 19 semi-structured expert interviews (Bogner & Menz, 2009) with geography teachers and extracurricular educators trained in ESD (see Table 3.1). All interviews were recorded, transcribed, and analyzed using qualitative content analysis according to Kuckartz and Rädiker (2023). We carried out a thorough web search for ESD learning places and schools with existing environmental projects. We specifically looked for extracurricular ESD learning places that had some form of mobility education in their portfolio and secondary schools that had been designated as Eco-Schools. We are aware that, due to the sampling method, the teachers and educators interviewed were probably already very motivated and interested in the field of ESD. Nevertheless, we are satisfied that the selection process can be considered good scientific practice for an exploratory study. Due to Germany's federal system, curriculum requirements vary from state to state, as each state publishes its own curriculum for each school subject (KMK, 2021). This has to be taken into account when doing research within the formal German education system, especially when doing so in different federal states.

Table 3.1 Sample structure of our study (own draft)

	Geography teachers	Extracurricular educators
Total (n = 19)	11	8
Age	27–54 years (Æ 38 years)	35–56 years (Æ 43 years)
Professional experience	1–22 years (Æ 10 years)	7–27 years (Æ 16 years)
Gender	8 female, 3 male, 0 other	2 female, 6 male, 0 other
Type of school/ organization	Comprehensive school (5), Grammar school (6)	NGO working in ESD (2), ESD learning center Lower Saxony (1), communal actors in mobility transition (3), freelancer in ESD (1), private company working in mobility education (1)

3.5 The Perspectives of Teachers and Educators on Mobility Transition

3.5.1 Comments of Geography Teachers and Extracurricular Educators on the Current Trends in (Sustainable) Mobility Transition

When asked for their definition of sustainable mobility, educators and teachers responded foremost with the ecological side of sustainability. Drastically reducing private car use for daily commutes was the first answer they gave. Instead, they advocated for the strong promotion of sustainable mobility alternatives (e.g., cycling) by political stakeholders.

In particular, the school teachers were convinced of their position as role models. By arriving at school via sustainable transportation themselves, they hoped to have a long-lasting impact on their students' mobility decisions. Vice versa, some teachers were afraid that by using their private car to get to school, they would lack the moral integrity to judge this behavior as bad in the classroom:

> … you can't sit there as a teacher and point your finger when all the students see me driving my car to school in the morning. That is of course a bit difficult (Teacher, I14_LK8_HH _0:01:58.0).

Extracurricular educators identified their role as change agents more through their learning provisions. They hoped to have a long-lasting impact on student behavior by engaging them in meaningful interaction and giving them the opportunity to critically consider possible alternatives. In addition, school teachers as well as extracurricular educators recognized an extremely strong influence of the students on their parents and vice versa, regarding their mobility decisions.

3.5.2 Reflections on the Digital Aspects of Mobility Transition and Its Social Implications

When asked to consider current smart mobility trends, teachers as well as educators assessed digital mobility as an inevitable part of progress. The examples most frequently named were bike/car sharing offers, autonomous driving, and digital infrastructure as part of smart cities (e.g., digital traffic light control). When asked about the relevance they would ascribe to the digitalization of mobility solutions, all interviewees rated it as very important. In particular, extracurricular educators highlighted that digitalization should not come at the expense of existing sustainable mobility options. Instead, they advocated for a thorough integration of old and new mobility solutions through digital means.

In the interviews, we asked the participants to share their thoughts on the social implications of smart mobility. As mentioned, most participants appraised the digital trends in mobility as good and necessary. When asked about the social implications of these developments, respondents raised two issues in particular. The first was the exclusion of the elderly or those unable to use a smartphone. Most respondents were convinced that analog and digital systems would need to coexist for some time to cater to those unable to use digital solutions:

> At the end of the day, I have to be careful that I take all groups with me. Especially those who perhaps CANNOT participate in digitalization in the way that others simply could (Extracurricular educator, I5_AM2_HH_0:37:30.3).

The second risk highlighted was data security. As digital solutions rely heavily on the collection, processing, and analysis of data provided by users, data breaches and the potential unethical use of the data were identified as potential negative impacts, particularly, because the data gathered included movement profiles, which were considered as highly sensitive.

Smart mobility was also associated with positive social implications. The most prominent aspects mentioned were the ability to switch languages within an app or to see which modes of transport are suitable for physically disabled people. Other positive aspects mentioned were smart ticketing and better availability of information for the hearing and visually impaired through the use of apps.

3.5.3 The Role of Digital Mobility Transition and Its Social Implications in Current Formal and Non-formal Education

Teachers and educators emphasized that digitalization itself is still a topic that does not play an appropriate role in current education in Germany. Although there are many links between digitalization and the main curricula in both states (Hamburg and Lower Saxony), there remains some uncertainty about where this topic can best be discussed. The majority of school teachers felt that geography education could be the appropriate subject to teach about the risks and benefits of digitalization because of the many possible links with the main curriculum. Other subjects, such as politics or computer science, were also seen as appropriate for discussing the more technical or ethical components of this issue.

Smart mobility is a small part within the broader topic of digitalization and was therefore considered particularly difficult to integrate into current teaching. Nevertheless, the school teachers interviewed identified some appropriate anchor points in the current core curriculum to discuss smart mobility and its social aspects (e.g., sustainable city development or regional disparities):

> Yes, we put that into context in the upper school with urban development. It's also partly about the topic of disparities, in a different semester. And the disparities semester is also about transport (Teacher, I11_LK7_HH_0:06:28.2).

When asked about the integration of the social aspects of mobility into teaching, the majority of teachers were very interested but had not dealt with the topic in a major way in their classes. Most mentioned the public discussion about the accessibility of transport during the COVID-19 pandemic as something they had raised in class.

With extracurricular educators, the offers vary greatly. Some educators offered a broad range of topics, such as smart mobility, and others were more focused on the classical parts of mobility education (e.g., bike repair workshops). It was mentioned that the classic mobility education workshops were also more frequently requested by schools. The social aspects of mobility did not play a key role in any of the extracurricular offers. All educators acknowledged that it was a blind spot in their offer and that the social aspects of mobility should be better integrated.

3.6 Discussion: Teaching Digital Mobility Transition and the Social Implications in Geography Education

School teachers and extracurricular educators are key actors in the context of ESD as they motivate young people to become change agents (UNESCO, 2020). By imparting knowledge and promoting young people's self-determined engagement, school teachers and extracurricular educators can enable young people to play a key role in a just and inclusive mobility transformation (Kanning & Meyer, 2022; UNESCO, 2017). However, we must acknowledge that educators as well as students "… feel inadequately prepared for their role as possible change agents" (Winter et al., 2022: 19), a conclusion supported by the results of our study. Teachers and extracurricular educators are well aware of the importance of the mobility transition as part of a larger push toward a sustainable future. We observed that the level of in-depth knowledge about smart mobility and its social implications is much higher among extracurricular educators than among school teachers. This seems unsurprising as extracurricular educators are professionals in the field of ESD and mobility education and therefore have much more work experience and knowledge on this specific topic. School teachers, on the other hand, have to deal with a variety of topics in the context of school geography education and are mostly constrained by curricular requirements. Although geography is positioning itself as the core school subject for ESD (Bagoly-Simó, 2023; Meadows, 2020), the hours available (usually one or two lessons per week in Germany) do not reflect this variety of topics. As the social aspects of smart mobility are only a sub-theme within the overarching topics of mobility transition and digitalization, it is up to the teachers to decide whether or not to integrate them. Mobility transition and smart mobility can be linked to a number of core topics in the current geography core curricula in

Hannover and Hamburg (e.g., regional disparities, sustainability, urban/rural development, and tourism). The topics of mobility transition and smart mobility are also very well suited for interdisciplinary learning (e.g., physics, chemistry, biology, physical education, computer science, or politics). However, this depends very much on the interests of the teachers and the willingness of the schools to allow other learning approaches.

One solution could be to integrate the social aspects of the digital mobility transition directly into the learning materials as the lack of such materials was mentioned by most of the teachers interviewed. Textbooks are updated only every few years and are often outdated. Additional materials are therefore essential to bring new empirical findings and scientific knowledge into school practice. In Germany, the available teacher's journals, with lesson suggestions for geography teachers are a good starting point for introducing digital mobility transition and its social aspects into current teaching. This is also true for the materials offered by extracurricular educators. Although they invest a considerable amount of time in developing materials on a variety of different topics, getting them to the schools and teachers was described as quite difficult. Unfortunately, we observed a general disconnect between extracurricular education offers and schools. Teachers were unaware of the extracurricular offers in Hannover or Hamburg, and extracurricular educators found it difficult to reach schools.

To address the concerns raised by the geography teachers and educators interviewed, we make three proposals (cf. Fig. 3.2):

1. There should be more opportunities for cross-curricular learning in schools. This could be achieved through project weeks, which are now an integral part of school education in Germany but would require the willingness of different school subjects to work together.

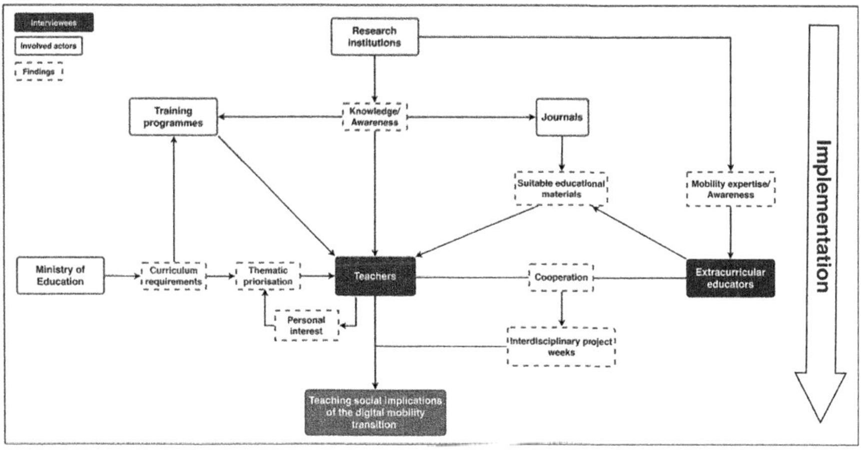

Fig. 3.2 Requirements for teaching the social implications of the digital mobility transition (own draft)

2. We advocate for a stronger link between extracurricular offers and schools. Geography teachers cannot cover all the details of the issues raised in ESD. They should, therefore, be able to rely on extracurricular experts to take some of the pressure off them. This could, in turn, lead to more detailed insights into important topics, as extracurricular educators are experts in their field and can usefully complement the geography curriculum.
3. Teacher training programs are needed to raise awareness of the social aspects of the digital mobility transition. As teachers occupy a central position in the school system, raising their awareness can subsequently lead to raising the awareness of many more students. These teacher training programs could either be organized either by extracurricular learning institutions or by universities, using the already existing teacher training infrastructure.

3.7 Conclusion and Outlook

Teaching the social implications of transitional change is an important aspect of achieving SDG 11 (UNESCO, 2017). As adaption to the looming climate crisis becomes more urgent, the discussion of proposed actions should always include all three dimensions of sustainable development, in addition to the political and cultural spheres (Meyer & Haubner, 2022). The results of our exploratory study show that school teachers and extracurricular educators do not cover the topic of smart mobility and its social implications in their lessons or educational offerings. We were, therefore, able to raise awareness of the social aspects of mobility among both groups. While teachers showed a high interest in integrating the topic into their lessons, extracurricular educators actively considered how to integrate the topic into their educational offers.

Based on our findings, we proposed three main recommendations:

1. Allowing more space for alternative learning formats, such as interdisciplinary project weeks, to address overarching societal trend
2. Strengthening links between schools and extra-curricular offers to complement the curriculum in geography education
3. Raising awareness of the social aspects of transformative change through teacher training programs and learning materials

We acknowledge that our findings are very specific to the German school system (and even to the states of Lower Saxony and Hamburg within Germany), as different countries have different curricula and requirements for geography education. Nevertheless, we are confident that our findings can be helpful in the context of ESD learning and the integration of new topics in geography education outside of Germany. Additionally, we have to consider, that our study was conducted in an urban environment and that the results might differ significantly in more rural school settings.

This research is funded by the Ministry of Science and Culture of Lower Saxony, Germany ("Niedersächsisches Vorab").

References

Axsen, J., Plötz, P., & Wolinetz, M. (2020). Crafting strong, integrated policy mixes for deep CO2 mitigation in road transport. *Nature Climate Change, 10*, 809–818. https://doi.org/10.1038/s41558-020-0877-y

Bagoly-Simó, P. (2023). Geography's unkept promises of education for sustainable development (ESD) on geography's wasted potential to educate for a more sustainable future. *International Research in Geographical and Environmental Education, 32*(1), 53–68. https://doi.org/10.1080/10382046.2023.2158631

Bogner, A., & Menz, W. (2009). The theory-generating expert interview: Epistemological interest, forms of knowledge, interaction. In A. Bogner, B. Littig, & W. Menz (Eds.), *Interviewing experts. Research methods series* (1st ed., pp. 43–80). Palgrave Macmillan.

Casadó, R. G., Golightly, D., Laing, K., Palacin, R., & Todd, L. (2020). Children, young people and mobility as a service: Opportunities and barriers for future mobility. *Transportation Research Interdisciplinary Perspectives, 4*, 1–11. https://doi.org/10.1016/j.trip.2020.100107

Dorsch, C. (2018). Reflecting on the smart city: How student teachers learn to teach smart pupils. *GI_Forum, 1*, 168–180. https://doi.org/10.1553/giscience2018_02_s168

Groth, S. (2019). Multimodal divide: Reproduction of transport poverty in smart mobility trends. *Transport Research Part A, 125*, 56–71. https://doi.org/10.1016/j.tra.2019.04.018

Hattie, J., & Zierer, K. (2019). *Visible learning insights*. Routledge.

IEA – International Energy Agency. (2023a). *Net zero roadmap: A global pathway to keep the 1.5 °C goal in reach*. https://www.iea.org/reports/net-zero-roadmap-a-global-pathway-to-keep-the-15-0c-goal-in-reach

IEA – International Energy Agency. (2023b). *Tracking Clean Energy Progress 2023*. https://www.iea.org/reports/tracking-clean-energy-progress-2023

IGU-CGE – International Geographical Union Commission on Geographical Education. (2016). *International Charta on Geographical Education*. https://www.igu-cge.org/wp-content/uploads/2019/03/IGU_2016_eng_ver25Feb2019.pdf

Jaramillo, P., Kahn Ribeiro, S., Newman, P., Dhar, S., Diemuodeke, O. E., Kajino, T., Lee, D. S., Nugroho, S. B., Ou, X., Hammer Strømman, A., & Whitehead, J. (2022). Transport. In P. R. Shukla, J. Skea, R. Slade, A. Al Khourdajie, R. van Diemen, D. McCollum, M. Pathak, S. Some, P. Vyas, R. Fradera, M. Belkacemi, A. Hasija, G. Lisboa, S. Luz, & J. Malley (Eds.), *IPCC 2022: Climate Change 2022: Mitigation of Climate Change. Contribution of Working Group III to the Sixth Assessment Report of the Intergovernmental Panel on Climate Change* (6th ed., pp. 1049–1160). Cambridge University Press. https://doi.org/10.1017/9781009157926.012

Kamargianni, M., Li, W., Matyas, M., & Schaefer, A. (2016). A critical review of new mobility services for urban transport. *Transportation Research Procedia, 14*, 3294–3303. https://doi.org/10.1016/j.trpro.2016.05.277

Kanning, H., & Meyer, C. (2022). The understanding and meaning of knowledge transfer for research and education. In M. Abassiharofteh, J. Baier, A. Göb, I. Thimm, A. Eberth, F. Knaps, L. Vilja, & F. Zebner (Eds.), *Spatial transformation. Processes, strategies, research design* (1st ed., pp. 9–26). Verlag der ARL.

Kenyon, S., Lyons, G., & Rafferty, J. (2002). Transport and social exclusion: Investigating the possibility of promoting inclusion through virtual mobility. *Journal of Transport Geography, 10*(3), 207–219. https://doi.org/10.1016/S0966-6923(02)00012-1

KMK – Secretariat of the Standing Conference of the Ministers of Education and Cultural Affairs of the Länder in the Federal Republic of Germany. (2021). *The Education System in the Federal Republic of Germany. 2019/2020. A description of the responsibilities, structures and developments in education policy for the exchange of information in Europe*. https://www.kmk.org/dokumentation-statistik/informationen-zum-deutschen-bildungssystem/dossier-englisch.html

Kuckartz, U., & Rädiker, S. (2023). *Qualitative content analysis: Methods, practice and software*. Sage.

Lucas, K. (2019). A new evolution for transport-related social exclusion research? *Journal of Transport Geography, 81*, Article 102529. https://doi.org/10.1016/j.jtrangeo.2019.102529

LSE – Leibniz School of Education. (2023, October 26). *Implementation in lower saxony: ESD decree, learning sites, model projects*. Retrieved August 8, 2024, from https://www.lse.uni-hannover.de/en/about-us/projects/teaching-change/translate-to-english-hintergrund-ueber-bne/translate-to-english-umsetzung-in-niedersachsen

Meadows, M. (2020). Geography education for sustainable development. *Geography and Sustainability, 1*(1), 88–92. https://doi.org/10.1016/j.geosus.2020.02.001

Meyer, C., & Haubner, R. (2022). "WIR reden MIT" – Jugendliche partizipieren an der Entwicklung von Lösungsansätzen für kommunale Zielkonflikte der Nachhaltigkeit. In J. Abt, L. Blecken, S. Bock, J. Diringer, & K. Fahrenkrug (Eds.), *Von Beteiligung über Kooperation bis zur Koproduktion* (1st ed., pp. 193–222). Springer VS.

Otto, I. M., Dongesa, J. F., Cremadesc, R., Bhowmik, A., Hewitte, R. J., Luchta, W., Rockström, J., Allerbergera, F., McCaffreyj, M., Doek, S. S. P., Lenfernal, A., Moránm, N., van Vuureno, D. P., & Schellnhuber, H. J. (2020). Social tipping dynamics for stabilizing Earth' s climate by 2050. *The Proceedings of the National Academy of Sciences (PNAS), 117*(5), 2354–2365. https://doi.org/10.1073/pnas.1900577117

Pangbourne, K., Mladenovic, M., Stead, D., & Milakis, D. (2020). Questioning mobility as a service: Unanticipated implications for society and governance. *Transport Research Part A, 131*, 35–49. https://doi.org/10.1016/j.tra.2019.09.033

Parkhurst, G., & Seedhouse, A. (2019). Will the 'smart mobility' revolution matter? In I. Docherty & J. Shaw (Eds.), *Transport matters* (1st ed., pp. 349–380). Bristol Policy Press.

Preston, J., & Rajé, F. (2007). Accessibility, mobility and transport-related social exclusion. *Journal of Transport Geography, 15*(3), 151–160. https://doi.org/10.1016/j.jtrangeo.2006.05.002

Scherf, C., & Wolter, F. (2016). *Electromobility. Overview, examples, approaches. Sustainable urban transport technical document #15*. Deutsche Gesellschaft für Internationale Zusammenarbeit (GIZ) GmbH.

SEU – Social Exclusion Unit. (2003). *Making the connections: Final report on transport and social exclusion*. https://www.ilo.org/media/312721/download

Sheller, M. (2020). Mobility justice. In M. Büscher, M. Freudendal-Pedersen, S. Kesselring, & N. Grauslund Kristensen (Eds.), *Handbook of research methods and applications for mobilities* (1st ed., pp. 11–20). Edward Elgar.

Smith, G., Sochor, J., & Karlsson, I. (2018). Mobility as a service: Development scenarios and implications for public transport. *Research in Transportation Economics, 69*, 592–599. https://doi.org/10.1016/j.rtbm.2018.12.003

UNECE – United Nations Economic Commission for Europe. (2020). *Transport trends and economics 2018-2019. Mobility as a service*. https://unece.org/DAM/trans/main/wp5/publications/Mobility_as_a_Service_Transport_Trends_and_Economics_2018-2019.pdf

UNESCO – United Nations Educational, Scientific and Cultural Organization. (2017). *Education for sustainable development goals. Learning objectives*. https://unesdoc.unesco.org/ark:/48223/pf0000247444

UNESCO – United Nations Educational, Scientific and Cultural Organization. (2020). *Education for sustainable development. A roadmap*. https://unesdoc.unesco.org/ark:/48223/pf0000374802

van Djik, J. (2020). *The digital divide*. Cambridge Polity Press.

Verlinghieri, E., & Schwanen, T. (2020). Transport and mobility justice: Evolving discussions. *Journal of Transport Geography, 87*(1), Article 102798. https://doi.org/10.1016/j.jtrangeo.2020.102798

Winter, V., Kranz, J., & Möller, A. (2022). Climate change education challenges from two different perspectives of change agents: Perceptions of school students and pre-service teachers. *Sustainability, 14*(10), Article 6081. https://doi.org/10.3390/su14106081

Jonas Koch is a geographer (M.Sc.), research assistant, and PhD candidate at the Institute of Science Education since August 2021. Since 2024, he has also been the coordinator for the joint project "Digitalisation for Sustainable Mobility" at the Green Office of Leibniz University Hannover. From 2018 to 2021, he worked as a consultant for sustainable mobility and urban transformation. His current research interests include sustainable mobility, mobility justice, urban transformation, and Education for Sustainable Development.

Tomke van Hove is an environmental planner (M.Sc.) who has been undertaking a traineeship in landscape management at the Lower Saxony Water Management, Coastal and Nature Protection Agency (NLWKN) since October 2024. From 2021 to 2024, she worked as a research assistant at the Institute of Science Education at Leibniz University Hannover. Her current research interests include Sustainable Development, biodiversity, and environmental law.

Christiane Meyer has held the Professorship for Geography Education at the Institute of Science Education at Leibniz University Hannover since 2008. From 1990 to 1998, she pursued studies and preparatory service for teaching at grammar schools (Gymnasien), followed by a year as a teacher from 1998 to 1999. Between 1999 and 2008, she worked as a research assistant and completed her PhD at the University of Trier. Her current research interests include Education for Sustainable Development, Transformative Learning, values education, and cultural awareness.

Chapter 4
Unveiling Environmental Complexity: Empowering High School Students Through Computer Simulations

Xi Xiang and Michael E. Meadows

4.1 Introduction

There are many complex and wicked problems facing the world today. Increasing population growth, urbanization, and the demand for natural resources have resulted in many complex environmental issues, such as anthropogenic climate change, land degradation, and loss of biodiversity (Kelly et al., 2020). Moreover, the suite of environmental issues is intricately intertwined, evolving through interconnections and feedback loops between multiple spatial–temporal scales to emerge as more serious issues. However, there is a lack of clear solutions to these issues because they involve abstract concepts and multiple dimensions, as well as uncertainty (Biermann et al., 2012). To address complex environmental issues and propose innovative solutions, it is essential to empower learners with advanced methodologies and tools (e.g., complex systems approach) for in-depth investigation.

X. Xiang (✉)
Independent Researcher, Shanghai, P. R. China
e-mail: shellyxiang1203@hotmail.com

M. E. Meadows
Nanjing University, Nanjing, P. R. China

University of Cape Town, Rondebosch, South Africa
e-mail: michael.meadows@uct.ac.za

M. Hardouin, C. Leininger-Frézal (eds.), *Curricular Changes in Geography Education*, International Perspectives on Geographical Education,
https://doi.org/10.1007/978-3-031-96551-7_4

4.1.1 *The Complex Systems Approach to Studying Interconnectedness*

Although environmental issues are extremely challenging to solve, they can be better understood when viewed through the lens of complex systems. A complex system is composed of many interconnected parts that interact with each other, leading to emergent behaviors that are not predictable from the properties of individual parts (Ladyman et al., 2013; Thurner et al., 2018). Studying the complex surface of the Earth is central to the discipline of geography (Cheng et al., 2022), and complex systems studied by geographers range widely from those dealing with the physical environment (e.g., climate, hydrology, and soils) to social and economic processes (e.g., urbanization and migration flows). Examining the interconnected parts of complex systems is helpful in unraveling the complexities of environmental issues. For example, the carbon and hydrological cycles can be envisaged as existing within a system framework that highlights how the Earth, atmosphere, and hydrosphere are interconnected. A complex system comprises different interconnected components that interact with each other, often in a nonlinear way, evolving through self-organization, for example, the effect of deforestation on climate at the global scale within a social-ecological systems framework (Mitchard, 2018). Therefore, complex systems analysis is an important approach to discovering the underlying mechanisms of human–environment systems and underpinning solutions for reducing and mitigating the negative impacts of complex issues (Turner & Baker, 2019).

4.1.2 *Systems Thinking in Geography Education*

To resolve the diverse challenges associated with environments and society, unique approaches to systems thinking have emerged within the discipline of geography that contribute to the understanding of complex systems in three ways. First, geography offers an integrative approach to consider the connections between human and natural systems, providing a foundation for studying these systems in depth. Second, the spatial context of a complex system is formed by the location and arrangement of its elements (Herman et al., 2024). Identifying complex patterns among these elements, which is enabled through a spatial perspective, advances system understanding, such as how some of the elements influence others. Third, as an attribute of spatial analysis, spatial scale, relating to the size of the area being studied, is an important consideration. The manifestation of complex systems at different scales, from local and regional to global, results in scale-dependent complexity (Cheng et al., 2022). This approach allows for appreciation of variability across areas within the system and the hierarchy between systems. By drawing on approaches from both complex systems and geography, we emphasize the value of complex systems in geography education, which guides the development of a new curriculum to address prevailing environmental challenges.

4.1.3 Fostering System Competency to Understand Complex Issues

Since system competency is a very important topic of geography education, it is necessary to include appropriate curricular content in the classroom and nurture student competencies to understand complex systems. System competency refers to the ability to identify, comprehend, and model complex and domain-specific reality (Lux & Budke, 2020). In comparison, complex systems competency engages students in analyzing these systems, which are frequently characterized by non-linear relationships. In the science curriculum, complex systems competencies involve determining system interdependencies and recognizing the emergent properties of the system (NGSS Lead States, 2013). In the subdomain of Earth and space science, students are required to explore not only the relationships among Earth systems but also how human activities modify these systems. In general, studying complex systems competencies, an integrated element in geography education, requires students to recognize and explain complex systems' structures, processes, and dynamics (Mehren & Rempfler, 2022).

4.1.4 Harnessing Technology to Enhance Learning About Complex Systems

System competency is introduced to students from the secondary level to the tertiary level to promote their understanding of complex environmental issues (Scherer et al., 2017; Cox et al., 2019). However, when introduced in the geography classroom, interactions and relationships among components within the system are given more attention (Bielik et al., 2023) than emergent behaviors, such as responses to perturbations, and self-organization. Additionally, among the variety of methods that can be used to learn about complex systems, computer-supported tools such as modeling platforms are increasingly being adopted by students to seek solutions that address critical sustainability challenges (Haas et al., 2020; Rates et al., 2016). Accordingly, here we identify a research problem that can be formulated as "How can a technology-supported curriculum be developed to enhance students' learning about complex systems associated with environmental challenges?" The following sections outline the development of a technology-supported curriculum within the overall school geography syllabus content. We then design a curriculum module on wildfire spread that nurtures complex systems competencies of high school students.

4.2 Theory-Based Conception of a Technology-Supported Curriculum

This section elaborates on the main features of complex systems and discusses what topics relating to complex systems are suitable to be taught to high school students in the geography classroom, and how this can be accomplished through the development of an innovative curriculum.

4.2.1 Elements and Characteristics of Complex Systems

Among the many features of complex systems, the three most important in relation to geography education are interconnectedness, nonlinearity and unpredictability, and effects (Mehren & Rempfler, 2022). To begin with, the multiple interconnected elements in a system interact with each other and are subject to feedback loops. For instance, the complex system of anthropogenic climate change includes multifaceted and reciprocal interactions in the coupled human–climate nexus (Min et al., 2024). More specifically, while climate is a vital context for human activity, humans play a key role in climate change. Second, the behaviors and patterns of the system exhibit nonlinearity and unpredictability, where the properties that emerge in a system level are not observable at subsystem levels (Jensen, 2023). In other words, changes in one part of the system may have disproportionate effects on other parts. Third, predicting possible consequences and complex effects of the system behavior is a deliberate effort to influence system behavior toward a desired goal (Mehren & Rempfler, 2022). For example, fire regulation and management efforts (e.g., creating fire breaks and removing dead biomass) can directly influence the behavior of the wildfire and alter the system dynamics.

4.2.2 Setting Learning Goals for Complex Systems in the School Geography Curriculum

The development of an innovative curriculum should consider key elements of the learning environment, such as the learning goals and outcomes, technology and digital tools, and classroom teaching methods (Organization for Economic Co-operation and Development, 2013). The content of complex systems is typically only implicit in the geography curriculum, instead being covered indirectly through a range of interconnected topics. For example, the geography curriculum in England requires high school students (in this case A-level) to grasp concepts in relation to complex systems, such as causality, systems, equilibrium, feedback, interdependence, mitigation, and adaptation (Department of Education, 2014). Additionally, learners are expected to be able to recognize and analyze the complexity of people–environment interactions across geographical scales. The Chinese high school geography curriculum emphasizes integrative thinking, which is similar to systems thinking, as one of the four key competencies (Ministry of Education of the People's Republic of China, 2018). This type of thinking supports students in recognizing the interconnectedness of geographic features, analyzing the interactions between people and nature, appreciating the dynamic features of geographical processes, and understanding the impacts of multiple factors on a region's geographic characteristics.

In connecting learning outcomes in the high school geography curriculum (Department of Education, 2014; Ministry of Education of the People's Republic of China, 2018) with features of complex systems (Bielik et al., 2023; Yoon et al., 2019), we highlight what is needed to improve learner's competency in relation to complex systems (Table 4.1). As complex systems competency can be regarded as a specialized aspect of the broader skill set encompassed by system competency, we align both types of competencies according to the dimensions of system competency revealed by Mehren and Rempfler (2022) and Roczen et al. (2021) (Table 4.1). Specifically, interconnectedness is a key factor in identifying spatial interactions between the physical environment and human activities, which signify the characteristics of the system organization dimension. Given the system behaviors dimension, simulating dynamic behaviors of geographical and environmental systems entails mastery of features such as nonlinearity, emergence, evolution, and feedback loops.

Table 4.1 Complex systems competencies for the high school geography classroom. Source: Adapted from Mehren and Rempfler (2022) and Roczen et al. (2021)

Complex systems competency	Complex systems feature	System competency in geography education
Identify interactions between individual components in geographical and environmental systems at multiple spatial scales	Interdependence	System organization
Analyze dynamic process of geographical and environmental systems over time	Non-linearity, emergence, self-organization, feedback and cycles, evolution	System behavior
Predict the effects of human actions on geographical and environmental systems.	Effects	System-adequate intention to act

4.2.3 *Selecting Computational Tools to Support Students' System Competency*

Research has indicated the difficulties that students experience in comprehending complex systems, in terms of recognizing the nonlinear relationships between variables, predicting the effects of human activities on the physical environment, and considering dynamic behaviors that evolve over time (Cox et al., 2019). Despite these difficulties, complex systems can be studied using a variety of digital tools, which lie at the interface of computer science, geography, and Earth science disciplines. Geospatial tools are able to capture and represent the inherently spatial nature of human-landscape systems and have thus been widely utilized to grasp the interconnectedness and interrelatedness of components of such systems (Houser et al., 2017; Saidi et al., 2021). For example, with the help of Geographic Information Systems (GIS), students may perform fire risk assessment and mapping and determine how topography, climate conditions, and human activities combine to trigger wildfires.

Complex systems have also been studied using computer simulations, thereby improving understanding of their structural characteristics and dynamic responses. Computer simulations are defined as computer programs that mimic real-world processes or systems with mathematical models (Rates et al., 2016). These interactive simulations are effective tools to convey dynamic system behaviors, which are difficult to implement in real-life situations, such as fluvial erosion, wildfire spread, and urbanization (Fisher et al., 2021). More significantly, they demonstrate the potential to support students' conceptual understanding of the components for complex systems in the environmental science classroom (Rates et al., 2016). For instance, simulations created by agent-based modeling platforms (e.g., StarLogo Nova) enable learners to predict the trajectory of the current status and explore future change in social-ecological systems (Haas et al., 2020; Yoon et al., 2018).

4.2.4 *Pedagogy for Teaching Complex Systems*

As approaches to solving complex environmental issues necessitate distinct problem-solving skills, we propose applying innovative pedagogical methods in the geography classroom to more effectively teach complex systems, including experiential learning (Falloon, 2019), modeling-based learning (Louca & Zacharia, 2012), and creative problem solving (CPS) (Treffinger et al., 2006). These pedagogical approaches are examined here in the context of complex system competencies.

4.2.4.1 Experiential Learning

Modeling with computer programs brings experimental components and engages students in active experimentation. During the learning process, students test their emerging theories through hands-on work and reflect on tangible experiences to create new knowledge about the intricate patterns of interactions within complex systems (Falloon, 2019; Kolb, 2014). For example, geospatial tools such as GIS encourage students to observe the spatial structure of variables in geographical and environmental systems and discern interactions between these variables, which may be structured in non-linear ways. While interacting with computer simulations, students can virtually experiment with variables and systematically identify complex realities within a system.

4.2.4.2 Modeling-Based Learning

Models can help scientists in simulating real-world phenomena and in visualizing complex systems. With the modeling-based learning approach, students could tackle complex issues through identifying problems, synthesizing information, building up models, testing ideas, and finding solutions. In particular, they can analyze system behaviors, such as feedback loops and emerging and evolutionary properties while constructing models (Louca & Zacharia, 2012). However, we suggest that teachers acknowledge the difficulties that students face during model construction, refinement, and validation (Constantinou et al., 2019), and provide scaffolding when necessary. For example, explaining the fundamental science behind the systems is essential for students to interpret how geographical systems function and evolve when validating models.

4.2.4.3 Creative Problem Solving

Conventional thinking inadequately deals with the complexity of environmental issues in the Anthropocene, while CPS may yield innovative solutions. The CPS approach relies on less rigidly structured problem-solving processes and aims to find open-ended solutions to complex problems (Treffinger et al., 2006). Although it is less commonly utilized in classrooms (Merrotsy, 2017), CPS can inspire students to generate more ideas and options when they tackle challenging questions. In particular, it enables students to make connections between seemingly unrelated components and predict how changes to the complex system might be influenced by interventions. It is, therefore, recommended that geography educators employ this type of pedagogy more frequently so that students can explore the unintended consequences of interventions and arrive at more creative solutions to meet environmental challenges.

4.3 Development of a Learning Exercise in the Context of Wildfire Spread

By synthesizing the theories of complex systems, geography, and educational science, this section presents a learning exercise in the context of natural hazards, which helps high school students develop a comprehensive understanding of wildfire spread and ultimately discover solutions to contain these fires. A technology-supported curriculum module is designed for high school geography content, in which students are able to manipulate variables within the system and visualize potential outcomes using an online interactive simulation.

4.3.1 *Background Information*

Wildfires are a naturally occurring phenomenon and play a significant role in maintaining many natural ecosystem types (Kelly et al., 2020). However, an increase in severe fire activity due to climate change has become a pressing issue in a range of biomes globally. The rapid growth of cities and towns in the areas close to wildlands has exacerbated wildfire problems (Fisher et al., 2021) and has prompted a growing interest in mitigation efforts through a better understanding of fire behavior and fire systems mechanisms. Wildfire spread exhibits complex system features, as not all wildfires behave the same, and their characteristics depend on several environmental factors, including dynamic weather conditions, fuel availability, and topography (Fisher et al., 2021).

4.3.2 *Learning Task Design*

The technology-supported curriculum module is crafted based on the innovative Earth science educational resources that have been designed by Concord Curriculum, a non-profit educational research and development organization. In the classroom, students are expected to complete activities in which they are challenged to explore the factors influencing the progression of wildfires, and the risks and possible solutions associated with such a natural hazard (Table 4.2). The exercise reflects the complexity of environmental challenges brought by wildfires, and each task is aligned with knowledge about environmental change, as well as complex systems competency.

Table 4.2 Learning task design (after wildfire risks & impacts module available at http://concord.org)

	Task description	Environmental knowledge	Complex systems competency
Task 1	How do the drought level, wind speed and direction, terrain, and vegetation influence the speed and extent of wildfire spread?	Understand how geographical processes interact with each other	Comprehend interactions among the variables within the fire spread system
Task 2	What is the level of risk to three towns from the spreading Concord Wildfire? Which town is at the highest risk if each town has a similar-sized population?	Elaborate how distinctive physical and human landscapes change over time	Simulate the dynamics within the fire spread system
Task 3	Whether wildfires can be contained by firelines and helitacks in extremely dry conditions?	Address spatial interactions between human and environment	Evaluate the effectiveness of strategies in fire adaptation and mitigation response

4.3.3 *Simulation Tools for Exploring Wildfire Dynamics*

As a user-driven computer simulation with student-controlled variables, Wildfire Explorer (WE, available at: https://wildfire.concord.org/) is an excellent tool for studying complex fire-spreading systems. The most appealing feature is that it merges wildfire spread calculations with visualizations of fire expanding over landscapes to model fire dynamics in the United States (Lord et al., 2024). With such a tool, student users are able to change the model's input parameters to display environmental conditions that influence fire spread and apply sparks to specific regions to simulate different ignition points, thereby running the model to visualize non-linear relationships and observe emergent behaviors of wildfires. Figure 4.1 is a screenshot illustrating the point at which the user selects the vegetation type and the drought level in each terrain zone. Students are also prompted to manage risks and impacts related to wildfires, for example, in exploring which localities are the most at risk, what resources and techniques are needed to contain the fire, and where to apply them (Table 4.3).

4.3.4 *Teaching Strategies for Wildfire Spread*

System competency is blended with problem-solving strategies to create pedagogies that enhance student learning of the fire spread system (Table 4.3). By completing the first and second tasks, the students investigate wildfire hazards through modeling-based learning with an experimental strategy. They are then exposed to CPS and inquiry-based learning to analyze the complexity of components in the third task, such as mitigation techniques.

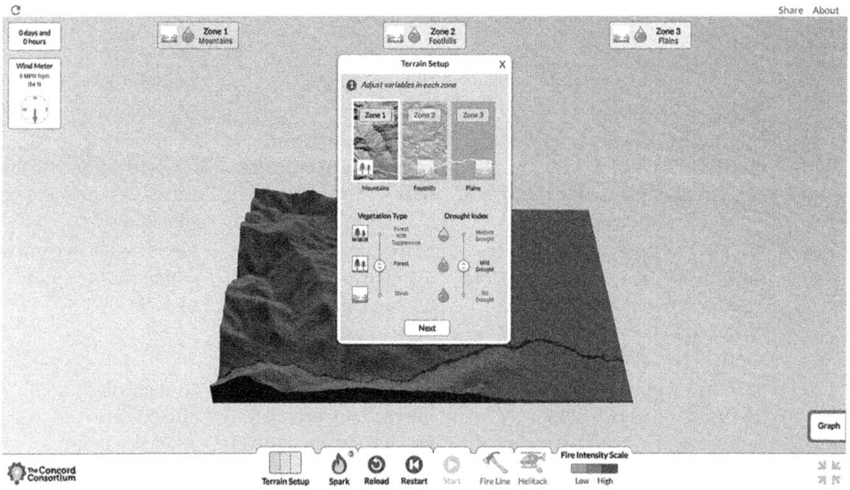

Fig. 4.1 "Forest" and "mild drought" are selected as vegetation and climate characteristics in the WE model (Source: Visualizations created based on Wildfire Explorer)

Table 4.3 Design of the technology-supported curriculum module

	Learning outcome	Technological functions	Pedagogical approach
Task 1	Understand the structure of the fire spread system	Visualize interactions	Experimentation, modeling
Task 2	Learn fire-spreading processes	Model dynamics	Experimentation, modeling
Task 3	Adapt with human actions	Predict the effects of intervention	CPS, inquiry-based learning

4.3.5 Implementation of the Learning Exercise

The first task encourages students to construct an understanding of how five variables of environmental conditions influence the spread and severity of wildfires. The purpose is to identify the interrelationships between individual components within the fire spread system. Having set the variables, including terrain, drought, weather and wind the same, students start to explore how vegetation type affects wildfire spreading behavior over time. By adding two ignition points in forest and grassland areas, they are able to visualize wildfire-spreading behavior across the two locations over time (Fig. 4.2). In making such a simultaneous, side-by-side comparison, students can understand the effect of vegetation type on wildfire spread. When all other variables in the model are equal, it becomes apparent that wildfires spread at different rates in grassland compared to forests,. Under teacher guidance, the students may then investigate underlying reasons while taking other factors into account, such as the fuel type, vegetation structure, size, and moisture level. It is

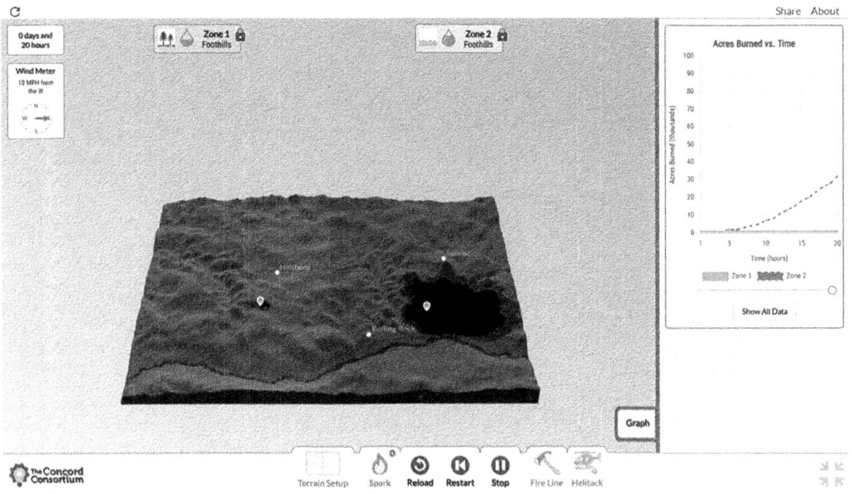

Fig. 4.2 Comparing wildfire spread in forest (left) and grassland (right) in the WE model (Source: Visualizations created based on Wildfire Explorer)

feasible to adopt the same logic to explore the effect of the other four variables on wildfire spread, thus identifying fundamental features of the system organization.

The second task involves comparing vulnerability to wildfires in different locations by analyzing system dynamics, such as where the wildfire is moving and its intensity, and determining which localities are at the highest risk. The level of risk is influenced by two factors: location relative to the fire and the associated environmental conditions. A hypothetical scenario is constructed in WE: viz. three towns, Peaksburg, Hillsboro, and Meadowland, which have similar population sizes but experience different levels of wildfire risk. The vegetation type is set as forest, shrub, and shrub for Zones 1–3, while the topography for Zones 1–3 is set as mountains, foothills, and plains (Fig. 4.3). By running models in WE, the students are able to visualize emergent properties of the wildfire and forecast that Hillsboro is exposed to highest levels of risk from the fire. After examining the environmental conditions in each terrain zone, the students understand that wind direction plays an important role in wildfire spread. If wind direction remains unchanged and the fire has not been contained, Meadowland is most impacted. Fire intensity there is higher than that in Hillsboro because the plains have flatter terrain than the foothills, offering less resistance to the continuous path of the fire. By the end of this task, students are in a position to discuss the implications of modeling wildfire spread, such as applying the results to determine appropriate evacuation routes for communities with early-warming systems in place. The modeling results also inform land-use decisions in the wildland–urban interface, that is, that urban and rural authorities should avoid development in areas that are especially prone to wildfire.

The third task provides insights as to how human interventions can produce changes in the wildfire spread system. Human response to wildfires, such as

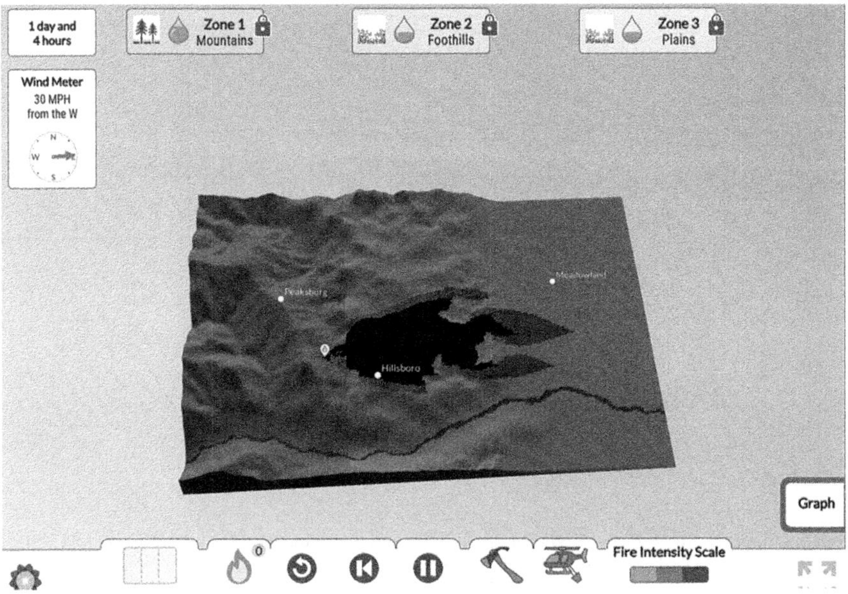

Fig. 4.3 Model wildfire risks experienced by three towns (Source: Visualizations created based on Wildfire Explorer)

firefighting efforts, directly influences fire behavior, thus altering the system dynamics. Firefighting efforts involve constructing a firebreak by clearing vegetation to create land strips and using helicopters that drop water to extinguish the flames. When running the model, students may find that a firebreak could contain the fire's spread and prevent it from reaching Skyview Town (Fig. 4.4). They may also discover that helicopters with water buckets are effective for protecting specific areas from fires and target smaller fires when the flames are not too intense. Geography teachers need to encourage students to compare multiple solutions that contain the fires and draw the conclusion that firebreaks provide a more strategic and long-term solution to wildfire risks as compared to helicopter water drops.

4.3.6 Assessment of Complex Systems Competencies

To evaluate which complex systems competencies are achieved, we recommend that students take a screenshot of the final state of the wildfire spread in the interactive simulations. They can then describe the characteristics of wildfire behavior, the level of wildfire risk, and the effectiveness of human interventions. This approach will demonstrate their mastery of intended learning outcomes, including system organization, system behaviors, and system-adequate intention to act, respectively.

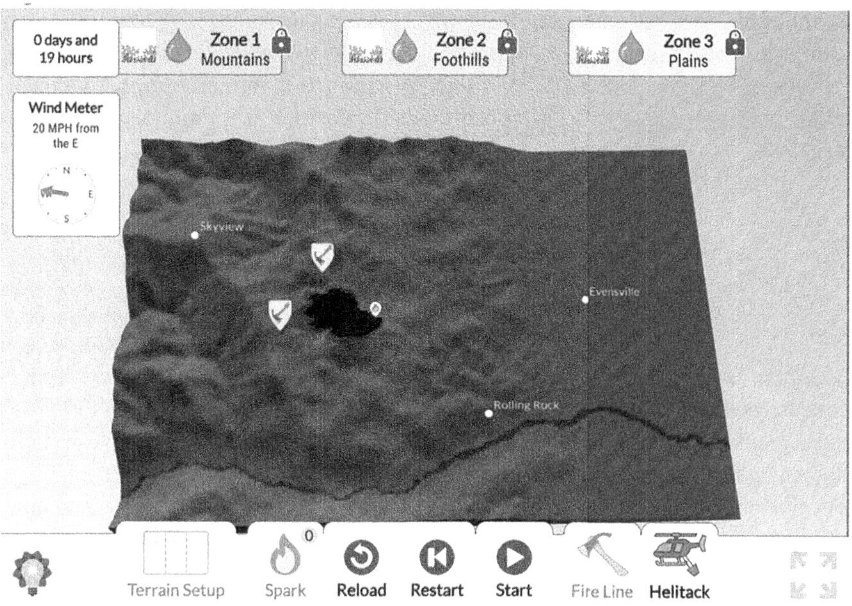

Fig. 4.4 Construct a firebreak to contain wildfire spread (Source: Visualizations created based on Wildfire Explorer)

4.4 Lessons from Designing the Technology-Supported Curriculum

This section discusses how adopting a complex systems approach contributes to understanding of complex environmental issues and in what ways integrating computer simulations into the technology-supported curriculum enhances students' engagement.

4.4.1 Understanding Wildfire Spread as a Complex Environmental Issue

This chapter applies a complex systems approach to investigate wildfire spread and assess fire risk. Different places may have varying levels of wildfire risk based on their locations and specific environmental conditions. In addition, the chapter emphasizes the link between the environment and society by highlighting human responses to ecological events like wildfires. On one hand, uncertainty around wildfire spread shapes the human response to this natural hazard, while, on the other hand, understanding the changes brought about by human activities is essential for comprehensively studying the dynamics of wildfires (Essen et al., 2023).

Furthermore, the learning exercise emphasizes spatial perspectives through motivating students to make more sustainable decisions that closely relate to real-world scenarios (McCaffrey et al., 2020), such as where to build home in the wildland–urban interface and how to establish communities that are more secure and fire resistant.

4.4.2 The Role of Computer Simulations in Studying Wildfire Spread

Based on existing literature and design experience from this research, we identify three ways that computer simulations such as WE support students' exploration and experimentation with complex systems. First, selecting from various options helps students gain knowledge about underlying mechanisms of the wildfire spread system (Fisher et al., 2021). By manipulating the variables in the model in an immersive manner, they can observe how wildfire spread is influenced by different environmental conditions and analyze the effects of wildfire risk on nearby settlements. Second, the immediate feedback that students receive from simulations makes it possible for them to predict emergent wildfire behaviors through altering individual components of the system (Lord et al., 2024; Rates et al., 2016). Third, as a simplification and abstraction of the interconnectedness of complex systems, computer simulations reduce the cognitive load for students, allowing them to focus on key variables and interactions of the system (Falloon, 2019). Overall, computer simulations in the classroom have great potential to promote a deeper understanding of complex system concepts through creating multifaceted scenarios, presenting a wide range of options, and prompting feedback.

4.5 Conclusion

The science of complex systems is a relatively new discipline, but the study of geographical and environmental systems is essential for solving complex issues, such as building a community more resilient to the risks posed by natural hazards. Accordingly, we have endeavored here to present a means of framing complex systems competencies in the school geography curriculum and developing an innovative computer-supported curriculum, in which high school students have the opportunity to learn the interactions, dynamics, and effects of interventions within complex systems under the support of computational tools and pedagogies. In the classroom exercise described, we demonstrate how student understanding of the complexity and emergent properties of the wildfire spread system can be positively facilitated through the use of simulations, in this case, WE. The importance of analyzing systems dynamics and the roles of human responses to complex systems is emphasized in the process.

Three areas for future research are proposed. First, there is a wide range of human–environmental problems and research needs to be conducted to explore to what extent students become aware of many other complex geographical systems, for example managing coastal development while preventing coastal erosion and sea level rise. Second, instead of using computer simulations based on pre-determined models, students may be guided to design their own systems through sourcing information, collecting data, and constructing models with agent-based modeling tools. Finally, we suggest further efforts be made to transfer state-of-the-art scientific research into educational materials in order to augment online learning resources that illustrate the impact of human activities on complex environmental systems.

References

Bielik, T., Delen, I., Krell, M., & Zvi Assaraf, O. B. (2023). Characterising the literature on the teaching and learning of system thinking and complexity in STEM education: A bibliometric analysis and research synthesis. *Journal for STEM Education Research, 6*, 199–231. https://doi.org/10.1007/s41979-023-00087-9

Biermann, F., et al. (2012). Navigating the anthropocene: Improving earth system governance. *Science, 335*, 1306–1307. https://doi.org/10.1126/science.1217255

Cheng, C., Cushman, S. A., Ho, H.-C., & Gao, P. (2022). Geographic complexity: Concepts, theories, and practices. *International Journal of Geo-Information, 11*(5), 308. https://doi.org/10.3390/ijgi11050308

Constantinou, C. P., Nicolaou, C. T., & Papaevripidou, M. (2019). A framework for modeling-based learning, teaching, and assessment. In A. Upmeier zu Belzen, D. Krüger, & J. van Driel (Eds.), *Towards a competence-based view on models and modeling in science education* (Vol. 12). Springer. https://doi.org/10.1007/978-3-030-30255-9_3

Cox, M., Elen, J., & Steegen, A. (2019). Systems thinking in geography: Can high school students do it? *International Research in Geographical and Environmental Education, 28*(1), 37–52. https://doi.org/10.1080/10382046.2017.1386413

Department of Education. (2014). *Geography GCE AS and A Level subject content*. https://assets.publishing.service.gov.uk/government/uploads/system/uploads/attachment_data/file/388857/GCE_AS_and_A_level_subject_content_for_geography.pdf.

Essen, M., McCaffrey, S., Abrams, J., & Paveglio, T. (2023). Improving wildfire management outcomes: Shifting the paradigm of wildfire from simple to complex risk. *Journal of Environmental Planning and Management, 66*(5), 909–927. https://doi.org/10.1080/09640568.2021.2007861

Falloon, G. (2019). Using simulations to teach young students science concepts: An experiential learning theoretical analysis. *Computers & Education, 135*, 138–159. https://doi.org/10.1016/j.compedu.2019.03.001

Fisher, R., Heckbert, S., & Garnett, S. (2021). Reframing wildfire simulations for understanding complex human–landscape interactions in cross-cultural contexts: A case study from northern Australia. *Fire, 4*, 46. https://doi.org/10.3390/fire4030046

Haas, A., Grapin, S. E., Wendel, D., Llosa, L., & Lee, O. (2020). How fifth-grade English learners engage in systems thinking using computational models. *Systems, 8*(4), 47. https://doi.org/10.3390/systems8040047

Herman, K., Ramlatchan, M., & Herman, R. (2024). In consideration of context: The potential of Geographic Information Systems (GIS). *TechTrends, 68*, 161–173. https://doi.org/10.1007/s11528-023-00916-z

Houser, C., Bishop, M. P., & Lemmons, K. (2017). Teaching complex concepts in the geosciences by integrating analytical reasoning with GIS. *Journal of Geoscience Education, 65*(3), 263–271. https://doi.org/10.5408/16-152.1

Jensen, H. J. (2023). *Complex science: The study of emergence*. Cambridge University Press.

Kelly, L. T., Giljohann, K. M., Duane, A., et al. (2020). Fire and biodiversity in the Anthropocene. *Science, 370*(6519). https://doi.org/10.1126/science.abb035

Kolb, D. A. (2014). *Experiential learning: Experience as the source of learning and development* (2nd ed.). Pearson FT Press.

Ladyman, J., Lambert, J., & Wiesner, K. (2013). What is a complex system? *European Journal for Philosophy of Science, 3*, 33–67. https://doi.org/10.1007/s13194-012-0056-8

Lord, T., Horwitz, P., Lee, H. S., Pallant, A., & Lore, C. (2024). Using simulations to support students' conceptual development related to wildfire hazards and Risks from an experiential learning perspective. *Journal of Science Education and Technology*. https://doi.org/10.1007/s10956-024-10126-8

Louca, L. T., & Zacharia, Z. C. (2012). Modeling-based learning in science education: Cognitive, metacognitive, social, material and epistemological contributions. *Educational Review, 64*(4), 471–492. https://doi.org/10.1080/00131911.2011.628748

Lux, J. D., & Budke, A. (2020). Playing with complex systems? The potential to gain geographical system competence through digital gaming. *Education Sciences, 10*(5), 130. https://doi.org/10.3390/educsci10050130

McCaffrey, S., McGee, T. K., Coughlan, M., & Tedim, F. (2020). Understanding wildfire mitigation and preparedness in the context of extreme wildfires and disasters: Social science contributions to understanding human response to wildfire. *Extreme Wildfire Events and Disasters*, 155–174. https://doi.org/10.1016/B978-0-12-815721-3.00008-4

Mehren, R., & Rempfler, A. (2022). Assessing systems thinking in geography. In T. Bourke, R. Mills, & R. Lane (Eds.), *Assessment in geographical education: An international perspective. Key challenges in geography*. Springer. https://doi.org/10.1007/978-3-030-95139-9_2

Merrotsy, P. (2017). *Pedagogy for creative problem solving* (1st ed.). Routledge.

Min, S., Morgenstern, O., & Sun, Y. (2024, October 20). Human influence on the climate system. IPCC Sixth Assessment Report: The Physical Science Basis. https://www.ipcc.ch/report/ar6/wg1/chapter/chapter-3/.

Ministry of Education of the People's Republic of China. (2018). *Geography curriculum standards for upper secondary schools* (2017 ed.). People's Education Press.

Mitchard, E. T. A. (2018). The tropical forest carbon cycle and climate change. *Nature, 559*, 527–534. https://doi.org/10.1038/s41586-018-0300-2

NGSS Lead States. (2013). *Next generation science standards: For states, by states*. The National Academies Press.

Organization for Economic Co-operation and Development [OECD]. (2013). *Innovative learning environments, educational research and innovation*. OECD. https://doi.org/10.1787/9789264203488-en

Rates, C. A., Mulvey, B. K., & Feldon, D. F. (2016). Promoting conceptual change for complex systems understanding: Outcomes of an agent-based participatory simulation. *Journal of Science Education and Technollogy, 25*, 610–627. https://doi.org/10.1007/s10956-016-9616-6

Roczen, N., Fischer, F., Fögele, J., Hartig, J., & Mehren, R. (2021). Measuring system competence in education for sustainable development. *Sustainability, 13*, 4932. https://doi.org/10.3390/su13094932

Saidi, S., Younes, A. B., & Anselme, B. (2021). A GIS-remote sensing approach for forest fire risk assessment: Case of Bizerte region, Tunisia. *Applied Geomatics, 13*(4), 587–603. https://doi.org/10.1007/s12518-021-00369-0

Scherer, H. H., Holder, L., & Herbert, B. (2017). Student learning of complex earth systems: Conceptual frameworks of earth systems and instructional design. *Journal of Geoscience Education, 65*(4), 473–489. https://doi.org/10.5408/16-208.1

Thurner, S., Klimek, P., & Hanel, R. (2018). *Introduction to the theory of complex systems*. Oxford University Press.

Treffinger, D. J., Isaksen, S. G., & Stead-Dorval, B. (2006). *Creative problem solving: An introduction* (4th ed.). Routledge.

Turner, J. R., & Baker, R. M. (2019). Complexity theory: An overview with potential applications for the social sciences. *Systems, 7*(1), 4. https://doi.org/10.3390/systems7010004

Yoon, S. A., Goh, S.-E., & Park, M. (2018). Teaching and learning about complex systems in K–12 science education: A review of empirical studies 1995–2015. *Review of Educational Research, 88*(2), 285–325. https://doi.org/10.3102/0034654317746090

Yoon, S. A., Goh, S.-E., & Yang, Z. (2019). Toward a learning progression of complex systems understanding. *Complicity: An International Journal of Complexity and Education, 16*(1), 1–19.

Xi Xiang obtained her PhD from Nanyang Technological University, Singapore. She was an assistant professor in geographic education at East China Normal University. Her research interests include developing spatial thinking skills through geospatial technology, environmental change research and education, key competencies in secondary geography textbooks, and learning analytics in online environments.

Michael E. Meadows received his PhD from the University of Cambridge and is currently a professor at Nanjing University. He is a Fellow of the International Science Council, the Royal Society of South Africa, the Royal Geographical Society, and a Foreign Fellow of the Chinese Geographical Society. He is past president of the International Geographical Union. His major academic achievements are in the fields of environmental evolution, the human-environment relationship in the Anthropocene, and sustainable development.

Chapter 5
Arguments in Geography Classes—Investigating Elementary School Students' Reasoning

Livia Reis Dantas de Souza and Sonia Maria Vanzella Castellar

5.1 Introduction

Ever since the National School Curriculum Base Nacional Comum Curricular (BNCC) was enacted, geographic reasoning[1] has been a significant theme of research in the area of geographic education in Brazil. Generally, the discussions revolve around two aspects; some do not give explicit definitions of geographic reasoning, while the other part considers aspects of reasoning derived from scientific knowledge that can be applied to the classroom with specific goals such as learning and deepening understanding of the world (Silva, 2021). When considering the necessity to establish in the classroom a model of thinking that is particular to Geography in order to equip students with a more critical and solid perception of socio-spatial phenomena, the most important query is how this can be implemented.

Science Education has been long emphasizing argumentation as a teaching methodology in its research pursuits for several decades (Duschl, 2007, 2008). Incorporating argumentation in teaching practices facilitates students to articulate their learning reasoning by means of interaction through communication and thinking processes with the acquisition of high-level skills (Jiménez-Aleixandre &

[1] We understand geographical reasoning as a particular way of producing an explanation of a specific socio-spatial phenomenon with elements that are particular to Geography. In the design proposed by Castellar and De Paula (2020), geographical reasoning involves gathering cognitive processes, spatial relations concepts, geographical principles and concepts, and means of spatial representation (maps) in order to comprehend a specific socio-spatial phenomenon (geographic situation).

L. R. D. de Souza · S. M. V. Castellar (✉)
FEUSP Universidade, Sao Paulo, Brazil
e-mail: livia.dantas@alumni.usp.br

M. Hardouin, C. Leininger-Frézal (eds.), *Curricular Changes in Geography Education*, International Perspectives on Geographical Education,
https://doi.org/10.1007/978-3-031-96551-7_5

Erduran, 2007). Based on the theory of argumentative reasoning (Mercier & Sperber, 2017), this methodology postulates that we use reasoning to both rationalize and elucidate. We argue to give explanations and evidence for our conclusions regarding any phenomena, both for ourselves and for others (which also makes it possible to evaluate someone else's argument). Furthermore, debating is an intrinsic part of the way scientific knowledge is generated (Duschl, 2007, 2008); thus, by bringing this methodology to Geography Education, gaining knowledge would encompass not only taking in scientific concepts and data but also comprehending scientific practice itself and the scientific way of structuring and constructing explanations and theories that are the object of study in schools.

Although Brazilian research on argumentative techniques in the teaching of Geography remains scarce and inadequately examined (Carli, 2023) in comparison to investigations on geographic reasoning, we can assume that there are important aspects to consider when exploring how Geography is learned, especially how one learns to reason geographically. Consequently, the concept of argumentation becomes available, as arguments reveal how reasoning was structured previously, indicating which theories (scientific or not) underpin it, what elements compose it, and how those elements are linked. By engaging in argument, we are privileged to gain an understanding of the construction of an idea and the thinking of students.

Therefore, the objective of this chapter was to analyze the arguments made by students during Geography classes regarding urbanization, migration, and housing conditions in the specific context of Cubatão, SP, Brazil, in order to identify similarities and differences between the reasoning used by students and those found in science that are valued by scientific literacy. In this way, we can explore further the elements that can promote understanding of the processes of learning and teaching Geography, as supported by written arguments that can reveal the way scientific and student reasoning about geographic topics was established.

5.2 Methodology

This research was conducted via a qualitative-interpretive analysis of argumentative texts written by senior elementary school students and an extract of a scientific production. In the context of the industrialization and urbanization processes of Cubatão, SP, Brazil, a problem based on a geographic situation guided the production of arguments as follows: *If some places in the city should be areas of environmental preservation, such as mangroves, for example, why should people occupy the mangrove?*

For comparison purposes, two types of texts were evaluated. We have started by evaluating an excerpt from Jutta Gutberlet's thesis, from the book "Cubatão: Development, Social Exclusion and Environmental Degradation" (Gutberlet, 1996: p. 71 and 73). From this first analysis, we sought to identify the epistemic practices associated with a scientific interpretation of a socio-spatial phenomenon by examining explanations validated by peers who answered the guiding question set out in

Table 5.1 Questions and explanatory models of Geography adapted from Harvey (1969)

Question	Explanatory model
How can the phenomenon under study be ordered and grouped?	Cognitive description: collection, ordering, and classification of data; theory is implicit
How is the phenomenon organized in terms of spatial form and structure?	Morphometric analysis: examination of shapes and patterns in space with a coordinate system
How was the phenomenon caused?	Cause and effect analysis: to study the factors that contribute to the occurrence of geographic distributions
How did the phenomenon originate and develop?	Temporal modes of explanation: it is a cause-and-effect explanation over a long period of time, it considers temporal changes in the distributions of geographic phenomena
How do particular phenomena relate and interact with the phenomenon in general?	Functional and ecological analysis: analysis of phenomena based on the role/function performed
How is the phenomenon organized into a coherent system?	Systems analysis: studies the general structure of the functions of interrelated parts

the previous paragraph. The analysis of the argument's Geographic content was based on Harvey's explanatory models of Geography (1969).[2] The epistemic practices correspond to questions as follows (Table 5.1):

The same problematic situation was crafted through discussions of Gutberlet's (1996) thesis and, subsequently, probed in a didactic sequence for a master's degree research involving around 100 senior year students (14 years old) at a public elementary school located in Cubatão, SP, Brazil, in 2018. Students' arguments production forms the basis of the second type of text, which can be found in this master's degree investigation (Souza, 2018) database. Through this second assessment, we sought to verify nuances in the students' arguments to compare them to validated scientific explanations of the phenomenon.

The comparison was drawn between the reasons, justifications, and other elements of the argument which appeared in both types of text. The form and the content were considered in order to visualize which structural and substantive elements of Geography composed the arguments. We emphasized content analysis, especially to identify the origin of ideas, whether scientific, ontological, cultural, and so on, that underpin student arguments and academic arguments.

[2] In this work, the author is interested in understanding the process of acquiring and codifying geographic knowledge, and additionally, establishing criteria for geographic arguments and inferences, to identify what makes them reasonable. When dealing with various explanatory models of Geography, which are either derived from methodology or empirical research, he brings forth a few non-exclusive and combinable possibilities. The choice of one or more explanatory forms would be linked to the type of question you want to answer.

To conduct the research, we used the *Cmaptools* software[3] and relied on argument structure models available from research that had already been carried out. Among them, we highlighted the argumentation evaluation scheme (Engelen & Budke, 2021) and mainly the Toulmin argument pattern (TAP) (Toulmin, 2003).[4]

5.3 Theoretical Support

5.3.1 Argumentation Models Used in the Analysis of Arguments Produced in the Classroom

We have searched articles relating to argumentative practices in Geography teaching and the most commonly accepted models used for evaluating argumentation. We have looked for research instruments used for analyzing the formal structure and the content/logic of arguments to assist our own methodology for analyzing student arguments and scientific (geographic) arguments.

Engelen and Budke (2021, 2022) are references to research on argumentation for classroom purposes in Geography. In their investigations, they have pursued to build the idea of what should be a *geographic argument.* Based on TAP, the geographic argument structure includes reasons to support an opinion, use of counter-arguments and rebuttals, exploration of multi-perspectives/points of view, and acknowledgment of spatiality and temporal relevance. Their criteria to evaluate the argument consider completeness, complexity, relevance to the problem, data accuracy from a scientific point of view, the validity of the connection between the claim and its justifications, multiperspective, and the indication of spatial references.

Other research on argumentation in Geography was undertaken by Harma et al. (2021), in which the analytical focus was on the form/structure of students' argument, disregarding the content or quality of the explanation or geographic argument that the students developed and presented.

---

[3] *Cmaptools* is a graphical tool to represent conceptual schemes and logical articulations of arguments.

[4] TAP is made up of the following: (1) *data* are facts that support the claim; (2) *qualifier* evaluates the strength of the data to make a claim (e.g., a certain conclusion would *always* or, *sometimes, frequently, possibly* occur); (3) *claim* is the conclusion or explanation; (4) *warrant* consists of justifications and statements that bridge data and conclusion; (5) *backing* endorses the warrant, achieved through the use of an accepted theory or ideas from a renowned author, for example; and (6) *rebuttal* stands for the limits of a claim or the counter argument.

Another contribution comes from the research of Davies (2002) who, despite not working specifically with argument models, provided important findings in the assessment of how explanations were constructed by students, contemplating what we were looking for, the geographical content of students' explanations. The author stresses the importance of constructing learning descriptors for assessment considering specific geographic content, such as (1) particularities of the area that encompasses heterogeneity in the study of places, (2) imagination's role in human actions in places, and (3) the subjective use of language in the description of these places. Regarding learning related to item 1, it would be achieved by knowing how to differentiate what is common to places and identifying what would be specific to each of them. As it pertains to items 2 and 3, the discussion focuses on the issue of subjectivity regarding the interpretation of geographic phenomena, which involves human imagination. The author seeks to clarify that it is necessary to have several models to enrich learning and to understand Geography, given that the element of imagination in human creation and the broad scope of possible interpretations of phenomena correspond to the diversity of people, their backgrounds, the paths they take, their language, and the numerous possibilities for communication.[5]

In general, the studies we have discussed focus more on the structure of the argument than its content. TAP is the main reference for analyzing arguments. Not much emphasis is placed on the validity of scientific arguments taught/learned in Geography classes, which could be correlated to an assessment of the content of the arguments. From these reflections, the issue of evaluating the content of the argument, which had been largely overlooked, imposed itself in our investigation. From Davies' findings, the following questions remained: are there any essential, non-negligible, and common elements in a so-called geographic argument scheme? What epistemological practices are related to this argumentative process in our area that need to be nurtured in the classroom to contribute to the foundation of a geographic argument?

5.3.2 *Epistemic Practices in School Geography*

By focusing on a teaching proposal that aims to witness the construction of knowledge in action, we support analyses of what happens in the classroom by establishing a relation with what occurs in the scientific processes of Academic Geography. Epistemic practices[6] (Kelly & Licona, 2018), discourses, and/or culture adopted in

[5] This idea reinforces the importance of considering students' *habitus* that appear on their explanations in the school context.

[6] According to Kelly and Licona (2018), epistemic practices constitute socially organized and interactively achieved ways in which members of a group propose, communicate, evaluate, and legitimize knowledge claims. They are interactional, contextual, intertextual, and consequential. These social processes have been routinized and standardized over time, considering the commonly accepted definition of knowledge, taking into account the evaluation of ideas in a public and historic context, and the importance of evaluating claims and explanations of knowledge by relevant groups.

a given scientific community were gauged by the rigorous examination of an academic text that supported the didactic activity proposed to elementary school students.

This proposal is centered on the importance of participation in cultural practices in the construction of knowledge. The focus is on the process (producing arguments from issues) and not on the product (reproducing scientific explanations), underlying the role of communication (discourses and language) in developing knowledge within learning communities. The contextual character is another crucial point in these practices, prompting us to reflect on what makes up the investigative process and the discourse that supports geographic explanation.

In this way, Bourdieu's concept of *habitus*[7] (2011a, 2011b) can be incorporated into the discussion, specifically regarding the ways of speaking and being characteristic of the culture of an epistemic field. Arguments and geographic reasoning that are produced by students are understood as culturally learned dispositions. The schemes of perception and appreciation, our understanding of the world, and even our relationship with the spaces we occupy are shaped through the *habitus* acquired socially, especially in our family and at school, and are related to our individual journey, the social position we occupy in the world, our backgrounds, and the different fields through which we have passed.

We perceive that argumentation, as a practice that we encourage in school, is embodied through the *habitus*, or, using another concept, on the epistemic practices of a given field. Reflecting on the examples to be discussed further in the text, if the question is of a more frequent or socio-scientific nature, one must ask: What will be the distinctive traits of *habitus* that will prevail in the arguments? Those of scientific explanations learned in school content or those related to *habitus* and the cultural capital that students bring from outside the school? We believe that, when we think about argumentation in schools, different *habitus* from diverse backgrounds come together. The *habitus* of the scientific field, necessary to be learned in the school environment, interacts with the school environment. The academic and cultural capital that the student brings with them into the classroom and the *habitus* resulting from experiences outside formal education make this equation more complex. The outcome that we aim to foster in geographic education is the ability to make arguments on socio-spatial themes that bring together multiple points of view and schemes of thought derived from various fields of knowledge in order to compose the argumentation of school geography.

Understanding the *habitus* through prior analysis of scientific arguments, one can identify elements of an academic *habitus* in Geography to deal with socio-spatial problems. This exercise allows us to differentiate more scholarly knowledge of academic geographic reasoning from geographic knowledge acquired through physical experience, more related to practical understanding. It is possible that the rationale based on scientific explanations and methods will be materialized in school activities, due to the fact that the *habitus* is fully realizable within the field in

[7] In Bourdieu's theory, *habitus* is a set of social dispositions that help us think, judge, and act in the world. This way of being, acting, and thinking equips us to live in the world and is acquired throughout socialization processes, through explicit learning or through frequency and custom.

which it was structured. However, can the learning acquired from socio-spatial experiences be assimilated and actually learned more effectively than what is provided through scientific/erudite knowledge at school? This is what we will be focusing on in the investigation.

5.4 Results and Discussion: Analysis of Scientific Arguments and Students' Arguments

Figures 5.1 and 5.2, which feature the arguments presented, were mainly created according to TAP, as most of the research did. We used the Cmaptools concept mapping tool to illustrate and analyze the argument on its formal basis. Content analysis was supported by discussions about models for examining arguments produced in the school context with characterizations of elements that constitute geographic reasoning. The pictures, therefore, show in a systematic way the elements of the students' arguments and the arguments in the academic text regarding its form/structure. The analysis of their distinct Geographic content for each of them was then further analyzed below.

Fig. 5.1 Argument structure of a scientific text based on TAP

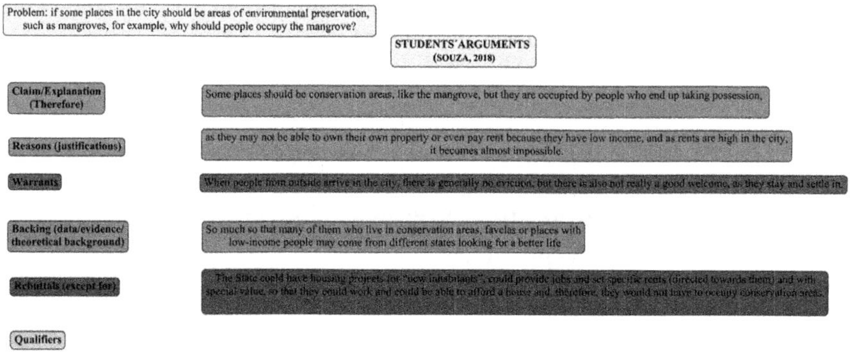

Fig. 5.2 Structure of the argument articulated in a text prepared by students based on TAP (Souza, 2024)

5.4.1 Evaluating the Argument in the Excerpt from Gutberlet's Book (1996)[8]

Regarding the structure of the argument, the researcher has included almost all elements of the TAP, excepting of the usage of qualifiers. The reasons are very coherently connected to the conclusion through sound evidence. To support her conclusion, the author frequently uses sources of theoretical principles and data from official data collection institutions. The author appeals to historical examples and to authorities on the subject to give more strength to her arguments. The incorporation of backing (data and references in scientific studies) is commonplace, and the section under analysis also includes a rebuttal, signifying the complexity of the argument being considered. These formal thoughts about the argument are more closely related to the amalgamated epistemic practices of the broader scientific field, not just Academic Geography.

Focusing on the content of the scientific argument in order to analyze evidence of characteristics of epistemic practices and habitus engaged in building the academic Geography argument, we relied upon David Harvey's (1969) book "Explanation in Geography." When comparing these precepts with the explanation produced by the scientific argument under analysis, we have: the author approaches the phenomenon by combining models of cause and effect, morphometric, functional, and temporal modes of explanation. It seeks to correlate cause and effect in the explanation by evaluating the spatial distribution of the housing type phenomenon with data focused on income and place of employment. Human geography and

[8] In 1991, Jutta Gutberlet's thesis was defended, and its content was later published in a book by EDUSP in 1996 titled "Cubatão: Development, Social Exclusion and Environmental Degradation." This book became a reference for studies involving the ecological and social impacts caused by the petrochemical and steel complex in Cubatão, SP.

physical geography are intertwined through the discussion of social elements (population data, age, occupation), territorial configuration (limited amount of land available for construction due to the presence of mangroves and mountains), and natural features (landforms, hydrography, biomes) in order to interpret and explain the reason for this arrangement/constitution/distribution of irregular and precarious housing configuration in the city. It mentions the functions of the city's areas and the historical process that culminated in the situation under study, using various scales (neighborhoods, city, region), comparing the income differences between neighborhoods, and comparing Cubatão to other cities in the region. Furthermore, geographic concepts and principles are used when explaining the location with precise directions, whether cardinal or the concept of urban centers and peripheries. From this content analysis, it is perceivable that epistemic practices expected within the Academic Geography field are present in the argument.

5.4.2 Assessment of Students' Arguments

The argument selected for analysis differs greatly in format compared to the scientific argument. There is a weak connection between reasons and conclusion through the warrant, by little recourse to the use of source-based backup, and finally by the lack of support from theoretical foundations or research that could reinforce claims.

Regarding its content, initially, the analysis was carried out according to the geographic reasoning proposal of Castellar and De Paula (2020). They indicate that when in a geographic situation, cognitive processes, concepts of spatial relationships, representations of space, geographic categories, and principles are interconnected, and this triggers geographic reasoning. Another possible front of analysis could be to consider geographic vocabulary used in students' arguments, with reference to concepts and terms typical to the area.

In the context of student argumentation, factual knowledge is prioritized over conceptual knowledge of the discipline, and geographic vocabulary is not common in written productions. Generally, the reasons provided to support the conclusion of their arguments are mostly from their own life experience, like the notion that income is a limiting factor when it comes to purchasing an appropriate place to live, or even the notion of how much rent costs in the city, which is closer to their experiences in the city.

Another issue regards the fact that students tend to draw comparisons between the place under study and different neighborhoods in the city, noting the infrastructure offered, that is, paying particular attention to observing the landscape, typical of what they can perceive in everyday life.

From Davies' investigation, we consider that this argument has indications of some characteristics of complex geographic content, such as the ability to differentiate places when comparing areas of the city that were and were not appropriate to live; by taking in multiple perspectives when including the State's role in solving the problem and considering the socioeconomic context on their solution; finally, by expressing their subjective imagination when proposing better job opportunities and dignity in housing projects.

The central question that remains after the analyses is whether the validity of this argument can be established. How can we assess whether the content of the argument demonstrates that the individual has acquired the capacity to reason geographically? Although we do not aim for our students to become geographers in school, we strive to help them broaden their understanding of the space around them. The structural organization and content elements of the scientific argument serve as a reference. Examining students' argument for correlations of cause and effect of phenomena, comparing the income level of different places in the city, and making an attempt to identify functions of each of these places could all reflect the reasoning and scientific understanding learned in the discipline. However, the evaluated production arises from a different context, the school context. For didactic purposes, it is necessary to be clear from the start about which learning objectives you intend to achieve through argumentation in the classroom. Would the goal be to provide students with scientific literacy, thereby enabling them to understand the step-by-step process involved in building knowledge in the field of Geography? Or would it be the broadening of student's repertoire, whether of their vocabulary or of skills and cognitive tools, to understand in greater depth their surroundings? Having several different models of explanations enhances students' abilities to understand socio-spatial phenomena, that is, we must intertwine models students bring from their family habitus and improve them with models derived from rigorous epistemic practices constructed in academic geography through time at school.

In evaluating these arguments, we must take into consideration the element of what is experienced by the students in their daily geography, what is not learned in the classroom, and what emerges from these situations in which they can express what they think, giving clues to the teacher about which individual explanatory models they already have on the topic. And these data cannot be overlooked, but rather taken into account and incorporated into the teacher's reflections on how to best help the student further their understanding.

Therefore, when teaching Geography, we must focus on the content and substance of the geographic explanation that students construct, rather than the form. And this implies the consideration of the perceptual experience one has of the phenomena studied, the ability to make coherent assertions and communicate this experience, as well as the role of students' geographic imagination.

In this context, we ponder whether the objective of geography classes is to reiterate established standards that have scientific validity as a learning goal, that is, the presence of so-called scientific truths and the academic modus operandi in argumentation as evidence of learning. Or if the objective is to focus on learning that enables students, rather than replicating models, to be able to argue based on a list of notions and reasoning provided by Geography so that they can develop a greater understanding of the world they live and the autonomy to think critically about the socio-spatial problems they come across. We believe that the second possibility would add more to students' education.

5.5 Final Considerations

This study has a limited scope, as it does not encompass a wide range of students' arguments, making it difficult to obtain more assertive conclusions about the argumentation process. In the master's research (Souza, 2018), with a greater number of participants involved, we saw an improvement in their ability to argue and conceptual expansion about their area of living. We believe that by exploring epistemic practices in the classroom, such as argumentation, we can foster the growth of students' geographic reasoning, enriching their thinking schemes on socio-spatial issues. During the process of producing arguments, students must put to test their ideas, justify them with evidence and theoretical background (concepts and principles), and contrast them to others' opinions, and, in this movement, new geographical vocabulary is better understood and incorporated, and collectively their mental schemes are refined. Learning is more effectively developed when we are able to set a scenario in which students are participants of the epistemic culture, when they are able to build their knowledge on the subject rather than just repeating descriptions and explanations that are readily available from a textbook.

From this initial comparison of students' and academic arguments, we believe that there is much that school Geography must achieve in order to facilitate geographical comprehension of socio-spatial phenomena. This includes primarily investigating what kind of reasoning is in students' minds when they interpret a certain geographical situation. As teachers foster students to argue in the classroom, we can have privileged access to their reasoning and can design strategies to help them gain more substantial geographical elements to their explanations. Epistemic practices can be learned in the classroom environment and may become students' habitus when they are required to think and discuss geographical situations whether in school or in other contexts.

Finally, a doctoral research is underway to create a more accurate argument evaluation tool in order to capture the nuances of the argumentative process and its content. Despite the vast and intricate field of argumentation studies and the limited amount of research done so far, we see that there are still important discussions concerning the process of teaching and learning Geography that can further the understandings achieved up to now.

References

Bourdieu, P. (2011a). Títulos e ascendência de nobreza cultural. In *A distinção: crítica social do julgamento* (2nd ed.). Zouk.

Bourdieu, P. (2011b). O habitus e o espaço dos estilos de vida. In *A distinção: crítica social do julgamento* (2nd ed.). Zouk.

Carli, E. (2023). *O modelo educativo AC-GEO (Argumentação científica a partir de um raciocínio geográfico) e a análise de seu efeito em promover mudanças conceituais* [Doctoral dissertation, Universidade Federal de São Paulo].

Castellar, S., & De Paula, I. (2020). O papel do pensamento espacial na construção do raciocínio geográfico. *Revista Brasileira de Educação em Geografia, 10*, 294–322.

Davies, P. (2002). Levels of attainment in geography. *Assessment in Education: Principles, Policy & Practice, 9*(2), 185–204. https://doi.org/10.1080/0969594022000001922

Duschl, R. (2007). Quality argumentation and epistemic criteria. In S. Erduran & M. P. Jiménez-Aleixandre (Eds.), *Argumentation in science education: Perspectives from classroom-based research*. Springer.

Duschl, R. (2008). Science education in three-part harmony: Balancing conceptual, epistemic, and social learning goals. *Review of Research in Education, 32*, 268–291.

Engelen, E., & Budke, A. (2022). Promoting geographic internet searches and subsequent argumentation using an open educational resource. *Computers and Education Open, 3*. https://doi.org/10.1016/j.caeo.2022.100090

Engelen, E., & Budke, A. (2021). Secondary school students' development of arguments for complex geographical conflicts using the internet. *Education Inquiry, 14*(1), 85–104. https://doi.org/10.1080/20004508.2021.1966887

Gutberlet, J. (1996). *Cubatão: Desenvolvimento, exclusão social e degradação ambiental*. Editora da Universidade de São Paulo.

Harma, K., Kärkkäinen, S., & Jeronen, E. (2021). The dramatic arc in the development of argumentation skills of upper secondary school students in geography education. *Education Sciences, 11*(734). https://doi.org/10.3390/educsci11110734

Harvey, D. (1969). Explanation in geography. .

Jiménez-Aleixandre, M. P., & Erduran, S. (2007). Argumentation in science education: An overview. In S. Erduran & M. P. Jiménez-Aleixandre (Eds.), *Argumentation in science education: Perspectives from classroom-based research*. Springer.

Kelly, G., & Licona, P. (2018). Epistemic practices and science education. In *Science: Philosophy, history and education* (pp. 139–165). Springer Nature. https://doi.org/10.1007/978-3-319-62616-1_5

Mercier, H., & Sperber, D. (2017). *The enigma of reason*. Harvard University Press.

Silva, P. (2021). *O raciocínio geográfico: Mobilizações intelectuais na interpretação de situações geográficas* [Doctoral dissertation, Universidade Federal de Minas Gerais].

Souza, L. R. D. (2018). *Potencialidades do conceito de território e da argumentação para a formação cidadã em aulas de geografia* [Master's thesis, Faculdade de Educação, Universidade de São Paulo]. https://doi.org/10.11606/D.48.2019.tde-11122018-092403

Toulmin, S. (2003). *The uses of argument*. Cambridge University Press.

Livia Reis Dantas de Souza is a PhD student at University of São Paulo (USP) and a Geography teacher in an elementary public school in Cubatão, SP, Brazil. Her research involves argumentation processes for learning geographical reasoning and her research interests are within the scope of argumentation, scientific literacy and Geography Education.

Sonia Maria Vanzella Castellar is a Full Professor at the Faculty of Education at the University of São Paulo (USP). She has a degree in Geography and her research focuses on school cartography, teaching methodologies, teachers´ training, scientific literacy and spatial thinking which are also her research interest topics.

Part II
Tools and Learning Approaches

Chapter 6
How Can Geographic Information Systems Increase Primary School Pupils' Awareness of Their Local

Elsa Filâtre

6.1 Introduction

Since 2016, French geography curricula for primary schools have encouraged teachers to base their instruction on pupils' neighborhood or local space, conducting field trips within it and considering it as lived space. In this chapter, the pupils' local neighborhood, corresponding to the space surrounding the school—as most pupils in France are enrolled near their homes—is often deemed familiar to them. However, research on children's mobility reveals a reduction in children's movement within their local space, more frequently confining them to a position of immobility and thus limiting their spatial awareness (Depeau, 2013; Filâtre, 2020). Moreover, the curricula focus on the concept of inhabit, which, beyond the idea of residing in a place, encompasses all practices within a space and how individuals interact with it (Lussault & Stock, 2010). This poses challenges for primary school teachers, who receive minimal training in geography, in both understanding and implementing these new curricula. The task, therefore, is to identify resources capable of helping pupils become aware of their neighborhood to evolve into inhabitants. Within the framework of a didactic engineering project conducted in 2018–2019 with various classes in the Toulouse region of southern France, the aim is to identify the stages that enable pupils to construct a geographical relationship with their immediate space (Thémines, 2016) and to demonstrate the extent to which Geographic Information Systems (GIS) can play a central role in this process. To observe the awareness-raising processes at work, multiple data sets were collected and analyzed during the implementation of a geographical survey in the pupils'

E. Filâtre (✉)
INSPÉ Toulouse Occitanie-Pyrénées, UMR GEODE, Toulouse, France
e-mail: elsa.filatre@univ-tlse2.fr
URL: https://geode.univ-tlse2.fr/

M. Hardouin, C. Leininger-Frézal (eds.), *Curricular Changes in Geography Education*, International Perspectives on Geographical Education,
https://doi.org/10.1007/978-3-031-96551-7_6

local space, taking their representations and questions as a starting point and relying on geographical visualization tools (GIS). The presented results will aim to demonstrate how GIS, in service of a geographical inquiry into pupils' local space, can serve as both a resource for teachers and as a lever for raising pupils' awareness of their immediate neighborhood.

6.2 Theoretical Framework: Spatial Awareness in Primary School Pupils

6.2.1 Geographic Information Systems: Opportunities in the Geography Classroom

School geography has a long tradition of using images of the Earth, in the form of maps, landscapes, films, satellite images, and, for over a decade now, geolocation tools and virtual globes. Geography is thus characterized by its multimodal means of expression (Thémines, 2020). Digital developments in iconography and cartography for both professional purposes, in the case of geomatics, and social purposes, in the case of geolocation and geovisualization, are now benefitting schools. Geomatics (contraction of the French terms for geography and computer science) can be defined as the use of digital tools to visualize, process, and communicate geographic information. It encompasses virtual globes (Google Maps, Géoportail, etc.), geolocation tools (e.g., Global Positioning System), and GIS.

Although Géoportail[1] is best placed in the category of virtual globe, it actually offers a complete ecosystem. It is a platform for knowledge about France's physical landscape that provides access to geographic reference information but has also become an infrastructure offering services to both the general public and professionals, in line with the European INSPIRE directive.[2] The platform enables data to be downloaded for professional GIS and includes applications that facilitate access to historical data on a given area, such as Going Back in Time.[3]

In schools, virtual globes and GIS are often confused and are often used more for location/situation tasks than for spatial analysis tasks, even though these tools have a strong didactic potential to build geographical citizenship conceptualized as a form of citizenship that enables individuals to engage with the world through the application of geographical tools and reasoning and to think and act spatially

[1] Géoportail: https://www.geoportail.gouv.fr/

[2] The European directive 2007/2/EC of March 14, 2007, known as the INSPIRE directive, aims to establish a geographic information infrastructure to promote environmental protection within the European Community. It was transposed into French law on October 21, 2010. https://www.geoinformations.developpement-durable.gouv.fr/directive-inspire-r296.html

[3] National Institute of Geographic and Forest Information (IGN)'s "Going Back in Time" application: https://remonterletemps.ign.fr/

(Genevois, 2008; De Miguel González & Lázaro, 2020; Mzuza & Westhuizen, 2019). This potential can be broken down into four forms of legitimacy (Fig. 6.1): social legitimacy, because the use of these tools helps to build pupils' geographical citizenship; technical legitimacy, because they reinforce digital skills; professional legitimacy, because they offer job opportunities; and above all pedagogical legitimacy (Genevois, 2008). The use of Géoportail and its "Going Back in Time" application offers genuine opportunities to facilitate pupils' geographical learning (Pawson, 2015; De Miguel González, 2021). Indeed, these tools not only allow pupils to engage in visual exploration and the discovery of space through different layers of spatial information (i.e., geovisualization) but can also become investigative tools for exploring and constructing approaches to spatial analysis.

In geography lessons in French primary schools, these tools can enable pupils to explore spaces visually, build itineraries, and answer questions they have about their living space.

6.2.2 *Framing Spatial Awareness in Primary School Pupils*

Recommended in classroom practice, and offering a strong didactic potential, the use of virtual globes can help pupils develop spatial awareness, especially of their immediate neighborhood, but the question is in what way? In the humanities and social sciences, the concept of critical consciousness (or conscientization) is rooted in research into awareness of oppression, notably in the work of Paulo Freire. The Brazilian educator explained that humans are situated and dated beings, and they must reflect and act upon their reality in order to become self-determining subjects rather than objects (Freire, 1965, 1970). In geography teaching, the focus is on identifying pupils' different levels of awareness in relation to objects of knowledge. We can identify three levels of awareness: awareness, awareness as intention, and awareness as knowledge, which focuses attention on the object of knowledge.

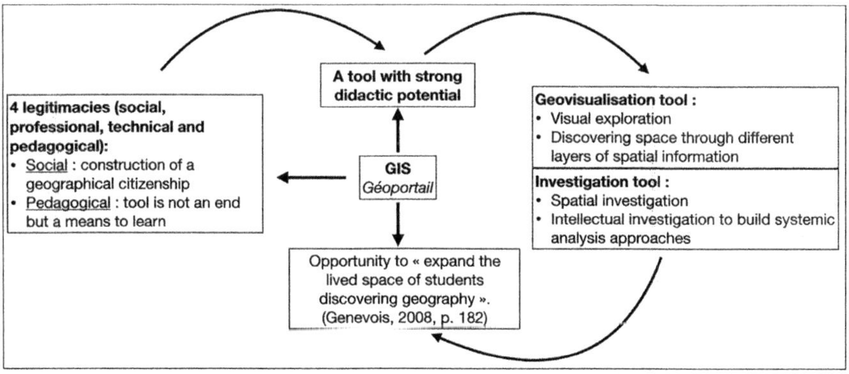

Fig. 6.1 Géoportail: what opportunities in the classroom?

Awareness can, therefore, take the form of a process whereby learners interact with the learning environment, enabling them to attend to an object, become aware of it, and then enter into a process of understanding (Fig. 6.2). The present research was based on the principle that space that is physically close to pupils is conducive to the construction of their spatial awareness.

6.2.3 Investigated Space: Building Pupils' Spatial Awareness

The didactic situation we tested in the classroom took the form of a geographical survey of the space close to the school, in order to be as close as possible to the pupils' lived experiences (Dewey, 1963; Kolb, 2014; Leininger-Frezal, 2018). This approach was developed according to the principles of place-based education (Israel, 2012; Pawson, 2015). First, in order to build on the pupils' experiences, we asked them to draw the space around their school and then to identify the questions they had about this space. After geographically categorizing the questions, the pupils thought about how to carry out the survey to answer them. The survey protocol was then put in place, based on various classroom practices: cartographic research using virtual globes (Géoportail and IGN's Going Back in Time application), real-life observation through field trips, interviews with local residents or councilors, statistical data research, family questionnaires, and documentary research. As the responses were formulated, the results were institutionalized to answer the central question of the survey: "How do we live in our neighbourhood?" At the end of the survey, pupils drew a new picture of the school neighborhood, to measure their cartographic learning (Filâtre, 2019).

The spatial awareness process, therefore, took place in several stages (Fig. 6.2).

Steps of the geographical survey of the space close to the school	Spatial awareness process

• **Starting from pupils' lived experiences :**
  • Drawing of the space around the school
  • Collection of pupils' questions about the space around their school
	Lived Space
	↓

• **Formalisation of the survey :**
  • Geographical categorization of pupils' questions
  • Research on the modalities of the geographical survey
	Questioned space
	↓

• **Setting up a survey protocol :**
  • Real-life observation throught field trips
  • Cartographic research using virtual globes (GIS)
  • Statistical data and documentary research
  • Interviews with local residents or councillors
	Investigated space
	↓

• **Formulation of answers by comparing results :** How do we live in our neighbourhood? **Evaluation of the sequence** (Drawing)	Spatial awareness

Fig. 6.2 Didactic situation designed to build spatial awareness

The local neighborhood is regarded as a lived space and therefore forms a link between spontaneous geography and school geography (Leininger-Frezal, 2018). We asked the pupils to question it so that they would attend to it and become aware of it as an object. By making it the focus of the investigation, we turned it into an investigated space that would allow pupils to develop several forms of spatial awareness. First, they would become aware that they were part of a reality (i.e., their local space) and that they were its subjects, as some social practices are spatialized. By becoming aware of their spatial nature, they would realize that they were doing with space (Lussault & Stock, 2010). Through encounters with other players, such as other pupils, parents, local residents, or local councilors, the pupils would discover that other people also engage in spatial practices in this space. And by observing the past and going back and forth between past and present using virtual globes, they would understand that their neighborhood space was not static, but part of a dynamic process.

Using visualization and investigative tools, our aim was, therefore, to help pupils turn their neighborhood space into an awareness space. The spatial awareness process would start with the pupils' own experience of the space. Through their investigation of this space, pupils would be able to construct it as an object and thus become aware of it as a shared and dynamic geographical space (Fig. 6.3).

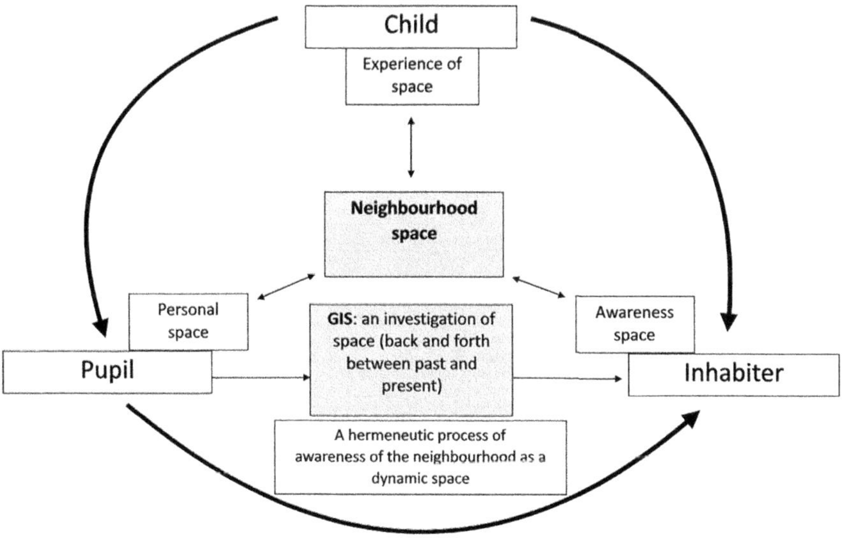

Fig. 6.3 Investigated space: building pupils' spatial awareness

6.3 Methods and Data Collection

6.3.1 Study Context

The present study was conducted within the framework of doctoral research (2017–2021) on how to develop the geographical awareness of pupils in the last 2 years of French primary school (9–10 years) using their neighborhood space. After administering an online questionnaire to CM1 (9–10 years) teachers in the Haute-Garonne department regarding their geography teaching practices and use of the local neighborhood, the objective was to work with the teachers of six classes[4] in schools across the Toulouse area (i.e., city center, city outskirts, and suburban and rural communities) on a didactic engineering project in the form of a geographical survey of the space around each school. This collaborative project between primary school educators and the didactic researcher took place during the first year of thesis fieldwork and, as shown in the diagram (Fig. 6.4), resulted in a booklet formalizing the survey. In the second year of thesis fieldwork, this booklet was sent to eight classes, again in schools across the Toulouse area.

6.3.2 Data Collection and Analysis

In order to observe the spatial awareness process, we collected and analyzed several sets of data. Table 6.1 summarizes the different stages of the data collection.

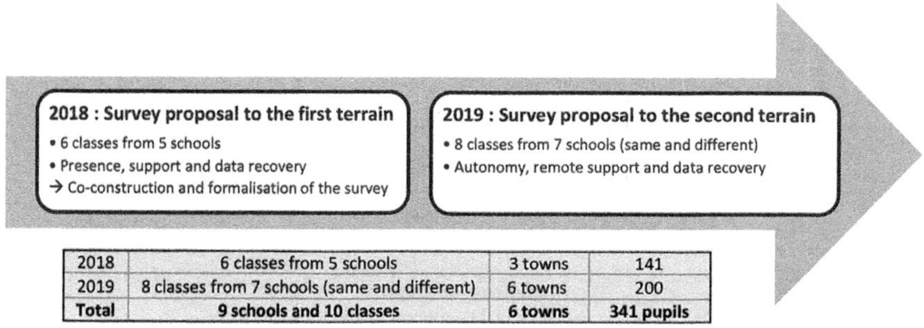

Fig. 6.4 Didactic engineering protocol

[4]The online questionnaire to CM1 teachers included a question to identify volunteers willing to participate in a research study on an investigative approach in the space around their school for teaching geography. Approximately 30 teachers responded to the call, and the selection was determined by their location and geography teaching practices to obtain a representative sample. The specific conditions of the study are detailed in the thesis (Filâtre, 2021).

Table 6.1 Different stages of the data collection

Stages and context	Data
2017: Online questionnaire administered to CM1 (9–10 years old) teachers in the Haute-Garonne department regarding geography teaching practices. A total of 147 responses were received and analyzed using the statistical analysis software (SPSS).	Online questionnaire on teaching practices, N=147 (SPSS analysis)
2018: Implementation of a geographical survey of the local area around schools, co-constructed between the researcher and six classes from different schools (urban, suburban, and rural) in Haute-Garonne. The survey resulted in the development of a booklet outlining the survey steps for training purposes.	Pupils, N=141 Video recordings of lessons, N=33 (Targeted transcriptions) Pupil interviews, N=52 (idem) Teacher interviews, N= 20 (idem)
2019: The survey was conducted in eight classes in Haute-Garonne without the researcher's presence. Learning outcomes were assessed using drawings representing the local area before and after the survey for all classes. (For results concerning the drawings, see Filâtre, 2021, 2023).	Pupils, N=200 Teacher interviews, N= 1 Online questionnaires on implementation of survey steps, N=8 (Data table)

6.4 Results: Geographic Information Systems, a Resource for Teaching Pupils About Their Local Neighborhood

6.4.1 Use of Geographic Information Systems Correlated with Outdoor Practice

The joint construction of the didactic engineering project was preceded by a statistical survey of the teachers' geography teaching practices, as we assumed that asking teachers who were not geography specialists to carry out a geographical survey with their pupils in the immediate vicinity of their school would be fraught with difficulty (Philippot, 2012). However, responses indicated that teachers viewed going out into the pupils' neighborhood as having a positive value and saw it as useful for geography learning. For 60% of the teachers questioned (N= 147), this practice corresponded to their idea of teaching geography. It was a good way of learning about geography because its concrete approach would help to motivate pupils (98% of responses, N=147). It was a popular way to work on finding one's bearings and asking questions about the neighborhood environment and provided a means of tackling geographical concepts through direct observation[5] (Filâtre, 2021). However, only one in two teachers interviewed had gone on a field trip to the local area to teach geography. One of the difficulties lay in designing resources about the local space, as teachers are used to following textbooks or methods that are spatially undifferentiated. Observing and questioning the local space means planning outings with pupils, designing itineraries and materials, and planning tasks and questions to

[5]The full results of the questionnaire are provided in the doctoral dissertation (pp. 203–239).

construct geographical concepts. These professional and didactic skills are far from self-evident and go some way to explaining why teachers find it difficult to use the neighborhood as a basis for learning geography.

The hypothesis that the didactic use of virtual globes has an impact on the teaching of geography was also probed in the questionnaire, according to the statements of the teachers interviewed. It emerged that Géoportail use was correlated with field trip practice.[6] Teachers who used Géoportail made up 41% of the sample, and 63% of these teachers went out into the neighborhood. It is, therefore, possible to conclude from these results that the use of virtual globes in the classroom is a lever for teaching pupils about the local space, at least insofar as it enables teachers to produce support materials (maps and plans) and construct itineraries.

6.4.2 Use of Geographic Information Systems for Geographical Survey

The geographical survey of the neighborhood took place in several stages, in which GIS were used in different ways. The first stage involved geovisualization. The pupils had to identify their school, their home, and all the places that had come to light as a result of their questioning. They then had to decide on an itinerary for their outing, which would provide answers to some of their questions. For example, their questions about the types of housing in their neighborhood led them to use the virtual globes to identify different types of residential space (individual or collective) in order to understand the residential characteristics of their neighborhood. Virtual globes were also used for spatial investigation. The Going Back in Time application could be used to answer a number of questions pupils might have about the development of their neighborhood. For example, fairly consistently across schools in all areas, pupils asked how long their school had existed, and whether certain facilities (shops, public services, etc.) had been there for a long time.

We examined this stage in the geographical survey by analyzing the transcripts of the class sessions. There were three possible stages in the use of the virtual globes: observation of maps to locate places in the neighborhood, construction of the itinerary for the outing, and analysis of the neighborhood in the past to answer questions about how it had changed. It emerged that, although not all the classes built the itinerary (only five of the classes carried out this stage), the Going Back in Time application was used by all eight classes. To build the itinerary, the teachers and pupils relied more on Géoportail's technical tools, which, therefore, exposed pupils to technical problems linked to the Internet connection. By contrast, the historical layers of the neighborhood could be explored either directly by the pupils or by the teacher, who could project the layers corresponding to the questions being asked. Teachers responded to three questions on a four-point scale about whether

[6] Correlation between Géoportail use and outdoor practice. Chi-square significant at 0.007.

the stage was easy or difficult to implement, from their point of view and that of the pupils, and about pupils' degree of motivation during this stage. Six out of eight teachers responded that the stage seemed fairly easy to implement, and all of them said it was easy for the pupils and motivating for them.[7] Generally speaking, all the stages based on virtual globes were perceived by the teachers to be highly motivating for the pupils.

6.4.3 *Geographic Information Systems as a Lever for Pupils' Spatial Awareness*

We now needed to identify the extent to which the use of virtual globes helped pupils enter into a spatial awareness process. We relied on two pieces of information: an extract from the transcript of a session using the Going Back in Time application and an extract from an interview with the class teacher.[8]

The pupils initially attempted to use Géoportail and the Going Back in Time application on their own, but, because of connection difficulties, the teacher organized a group session to enable everyone to see the different historical layers. The pupils had a double-entry table in front of them featuring nearby places such as the school, forest, lake, and old farm (rows), and maps/plans or aerial photographs from the historical layers (columns).

The teacher first displayed an aerial photograph and the IGN[9] map and then zoomed in with the help of the pupils, looking for landmarks that would help them find the neighborhood and the school. The pupils were very enthusiastic.

Observation of contemporary aerial photograph

Teacher: Can you see the forest?
Pupils: Yes!!!
Teacher: Can you see the school?
Pupils: Geography is so cool!

Observation of aerial photograph from the 1950s

Teacher: Did the school already exist? In my opinion, the school didn't exist because we saw that there were only fields! So we can put "no"?
Pupils: And the forest?
Pupils: It's there!
Teacher: And the lake?
Pupils: No!

[7] For six out of eight teachers, the pupils were highly motivated during this stage of the survey, and for the remaining two, the pupils were motivated. For full results, see doctoral dissertation (Filâtre, 2021, pp. 241–267).

[8] The description of the session featuring the use of the IGN's Going Back in Time application and the extract from the interview with the teacher concern the Suburban 2 class (Year 1).

[9] IGN: National Geographic Institute: https://www.ign.fr/

Observation of mid-nineteenth-century map

One of the pupils explained that if the school did not exist in 1950, it could not have existed before then.

Teacher: We'll go and check.
Pupils: There was no school. The lake didn't exist. The forest did. "La Paderne" too. Back then there were fields.

Observation of late-eighteenth-century map

Pupils: There was no school. The forest we know was already there. And you can't see the lake, it didn't exist.

Fresh observation of aerial photograph from the 1950s

Pupils: Forests, fields, farms, fewer houses, a central lane. Were there farmers in the fields?
Teacher: Look at the shape of the fields!
Pupil: I don't think it's natural because the lines are very straight.
Teacher: Compared with the forest. It's a manmade landscape. So how did people live before?
Pupils: Now we go to the supermarket, before they farmed in their fields.
Teacher: What are green spaces used for now?
Pupils: To breathe, to walk, to play. In the 1950s, the people who had fields used them to earn money.
Teacher: And what do we do for a living?
Pupils: In offices, in businesses, in buildings, in companies, in shops.

This transcript enabled us to identify the process of geographical reasoning that the pupils put in place. First, they set about locating the various listed places in the different historical layers, which led them in the final exchange to identify past practices, enabling them to deduce how these practices were spatialized. The school is located in a fairly recent suburban area and only featured in recent documents (i.e., aerial photographs taken in 2000). The pupils, therefore, deduced that the area occupied by the school was not a residential place, but a place dotted with farms and fields. They then deduced that these were not natural places, but places shaped by humans, who used them for farming. These practices could, therefore, be seen in the space (farms, fields), and had also left traces (the name of the neighborhood is the name of a former farming estate). The pupils eventually came to understand that the space they were observing allowed them to identify the practices of its inhabitants, and they were then able to compare these practices with those of people who had lived there in the 1950s. This geographical reasoning, made possible by spatial analysis, enabled pupils to see their neighborhood environment not as a static space, but as the result of a dynamic process of change. This gradual process of discovery was made possible by asking pupils to observe the different historical layers and by seeking explanations from the whole class.

In a Year 1 interview, one of the teachers highlighted the added value of the geographical survey in terms of pupils' understanding of space. When asked about how the survey of the local neighborhood contributed to the teaching of geography, she first identified a form of opacity for the pupils. As she put it,[10] "I didn't think we could do so much... er... and what's more... er... I didn't realise how difficult it was for children to position themselves and how important it was to connect them with neighbourhood, because I naively thought they could master the neighbourhood and that it would perhaps be a bit boring for them... well, not at all and we discovered a lot of things and I realized that there were things that hadn't been built up and building them from the neighbourhood... er... it wasn't easy and so I can't even imagine what it's like when it's based on a city they don't know, like when we want to study the map of Lyon, the map of Marseille or what can be done, or there they are... well... they imagine things but these can't be concrete and even when they are concrete it's hard for them."

The teacher went on to explain that the work on the geographical survey had made it possible to understand that geography is not just the study of locations and that, by working on their local neighborhood, pupils can become aware not only of their position as spatial actors but also of the practices of others and the dynamic process of inhabiting. She explained that "they were able to see that geography wasn't just about studying a map and that you could... that it was also about observing the way people live in their environment, so I don't think they had realized that, and the fact that they were able to move around, meet people, interview people, made them realize that they were actors and that this was something that moved... that's all. So that was one of the strongest contributions... the fact that a neighbourhood is alive, is dynamic, we live there, so as we live there we have an impact and ultimately that's what geography is all about, studying the place where we live and the impact that humans or the environment can have on space."

Thus, the geographical investigation of the local neighborhood relied on the use of GIS both for geovisualization of the space around their school and for constructing the initial elements of spatial analysis. Their utilization, coupled with other steps of the survey, particularly field trip, enables pupils' to engage in a spatial awareness process.

6.5 Discussion and Conclusion

Ultimately, the use of GIS in the classroom in the form of virtual globes (e.g., Géoportail and its dedicated historical application) makes it possible to initiate spatial awareness among pupils who are just beginning to learn geography. Our results showed that Géoportail can be an important resource for teaching geography. First, it is a resource for teachers who are not geography specialists, enabling them to find the materials, maps, and plans needed for place-based teaching. Second, it is a resource for the class as a whole, who can find the tools they need to carry out their

[10] Extract from the transcript of the interview with the teacher of the "Suburban 3" class (Year 1).

geographical surveys in response to their initial questions, to identify places, actors, and their practices, as well as the dynamics of territorial evolution through past–present comparisons. Although computer and digital equipment vary greatly between schools, with the studied classes being representative in this regard, the use of Géoportail, even when directed by the teacher without pupils' direct manipulation, produced the same effects in all studied classes. Third, as we have seen, it is a resource for individual pupils, who can use it to explore the world for themselves. The pupils were so enthusiastic that, at the request of the teachers, we produced a methodological guide sheet for them to use at home (contingent upon internet access and digital literacy).

Géoportail is an important tool for geography teaching, as it enables spatial citizenship to be constructed in the classroom by giving pupils an interest in their local area and its actors. It also gives them access to geovisualization tools and the ability to master the techniques involved. However, the most interesting contribution of virtual globes in the geography classroom is didactic, in that their use enables pupils to embark on a process of spatial awareness of their neighborhood, whereby they stop seeing it as something static and instead view it as the result of a dynamic process spanning the past, present, and future.

Finally, the use of virtual globes also shows that, contrary to what teachers often think, pupils are not familiar with the space in which they live. We intuitively assume that pupils' immediate environment is the one that is most familiar to them, but the assumption that pupils have experiential knowledge of their neighborhood has proved to be a major bias in the teaching of geography at school. Instead, research has revealed the opacity of pupils' everyday space, which can be explained by the declining place of children in the city and their reduced autonomy in their living space (Filâtre, 2020; Fotel & Thomsen, 2003; McMillan, 2005; Prezza et al., 2005). This makes it all the more necessary for schools to consider the local neighborhood as a relevant resource for teaching geography skills and building spatial awareness.

References

De Miguel González, R. (2021). Geographical and geospatial competences from school education to higher education: The contribution of international journals and EUROGEO in the international projects. *J-READING Journal of Reasearch and Didatics in Geography, 2*. http://www.j--reading.org/index.php/geography/article/view/299

De Miguel González, R., & Lázaro, M. (2020). WebGIS implementation and effectiveness in secondary education using the digital atlas for schools. *Journal of Geography, 119*, 1–12. https://doi.org/10.1080/00221341.2020.1726991

Depeau, S. (2013). Mobilité des enfants et des jeunes sous conditions d'immobilité? Ce que circuler signifie. *e-Migrinter, 11*, 103–115. https://doi.org/10.4000/e-migrinter.302

Dewey, J. (1963). *Experience and education*. Scribner Paper Fiction.

Filâtre, E. (2019). *Enquêter sur l'environnement proche de l'école avec ses élèves. Géographie CM1*. https://hal.archives-ouvertes.fr/hal-03614214

Filâtre, E. (2020). Les trajets domicile-école des élèves de primaire peuvent-ils constituer un point de départ pour construire des apprentissages spatiaux ? *Géocarrefour, 94*, Article 94. https://doi.org/10.4000/geocarrefour.14769

Filâtre, E. (2021). *Développer la conscience géographique des élèves en enseignant à partir de l'espace proche* [Phdthesis, Université Toulouse le Mirail - Toulouse II]. https://tel.archives-ouvertes.fr/tel-03560832

Filâtre, E. (2023). Les dessins de représentation de l'espace d'élèves d'école primaire : Une trace des apprentissages géographiques des élèves. *Recherches en Didactiques, 35*(1), 97–119. https://doi.org/10.3917/rdid1.035.0097

Fotel, T., & Thomsen, T. U. (2003). The surveillance of children's mobility. *Surveillance & Society, 1*(4), 10.24908/ss.v1i4.3335.

Freire, P. (1965). *Éducation et conscientisation.* http://acervo.paulofreire.org:8080/jspui/bitstream/7891/4657/2/FPF_OPF_01_0029.pdf

Freire, P. (1970). *Pedagogy of the oppressed* (30th anniversary ed.). Continuum.

Genevois, S. (2008). *Quand la géomatique rentre en classe.
Usages cartographiques et nouvelle éducation géographique dans l'enseignement secondaire* [Phdthesis, Université Jean Monnet - Saint-Etienne]. https://tel.archives-ouvertes.fr/tel-00349413/document

Israel, A. (2012). Putting geography education into place: What geography educators can learn from place-based education, and vice versa. *Journal of Geography - J GEOGR, 111*, 76–81. https://doi.org/10.1080/00221341.2011.583264

Kolb, D. A. (2014). *Experiential learning: Experience as the source of learning and development.* FT Press.

Leininger-Frezal, C. (2018). Training primary teachers through experiential geography. *European Journal of Geography, 9*(2), 17.

Lussault, M., & Stock, M. (2010). "Doing with space": Towards a pragmatics of space. *Social Geography, 5*, 11–19.

McMillan, T. E. (2005). Urban form and a child's trip to school: The current literature and a framework for future research. *Journal of Planning Literature, 19*(4), 440–456. https://doi.org/10.1177/0885412204274173

Mzuza, M. K., & Westhuizen, C. V. D. (2019). Skills gained ant their significance in teaching with or through GIS/EJG. *European Journal of Geography, 10*(3) https://eurogeojournal.eu/showPaper.php?id=1544

Pawson, E. (2015). What sort of geographical education for the anthropocene? *Geographical Research, 53*(3), 306–312. https://doi.org/10.1111/1745-5871.12122

Philippot, T. (2012). Enseigner à l'école primaire une géographie problématisée : Un défi? *Nouveaux c@hiers de la recherche en éducation, 15*(1), 21. https://doi.org/10.7202/1013377ar

Prezza, M., Alparone, F. R., Cristallo, C., & Luigi, S. (2005). Parental perception of social risk and of positive potentiality of outdoor autonomy for children: The development of two instruments. *Journal of Environmental Psychology, 25*(4), 437–453. https://doi.org/10.1016/j.jenvp.2005.12.002

Thémines, J.-F. (2016). Propositions pour un programme d'agir spatial : La didactique de la géographie à l'épreuve de changements curriculaires. *Les Sciences de l'éducation - Pour l'Ère nouvelle, 49*(4), 117. https://doi.org/10.3917/lsdle.494.0117

Thémines, J.-F. (2020). Comment des élèves lisent un document composite en géographie. *Pratiques. Linguistique, littérature, didactique*, Article 185–186. https://doi.org/10.4000/pratiques.8416

Elsa Filâtre is an Associate Professor at the GEODE laboratory, specializing in Environmental Geography at the University of Toulouse Jean-Jaurès and the INSPE (Teacher Training Institute) in Toulouse. Since completing her PhD in geography didactics in 2021, she has been working in the field of geography didactics. Her research focuses on pupils' spatial representation drawings as a means to analyze children's spatiality and explore student participation processes in the context of geographical learning and education for sustainable development.

Chapter 7
Multi-sensory Developments in Geography Education. A Historical Analysis of the Situation in French Educational Institutions

Jean-Pierre Chevalier and Guilhem Labinal

7.1 Introduction: Research Question and Protocol

Multisensory approaches, increasingly advocated in geography education, challenge three aspects of the discipline: epistemological, methodological, and scalar. Epistemologically, they reject the decoupling between object and subject, that is, between space and the individual who perceives, inhabits, and senses it. Methodologically, they question the importance of visual practices in contemporary geography teaching. Finally scalar, by granting sufficient significance to the study or the relationship established by the individual with space to shed light on the relationship that a group, a society maintains with it.

Several research projects take these approaches as their object of study in geography didactics. In didactic research, some studies focus on these approaches as objects of study, but considering the diversity of pedagogical initiatives that mobilize them to teach geography in and outside the classroom, it remains necessary to identify the legacies of previous educational practices. This chapter, therefore, proposes to restore, from a historiographical perspective, previous practices by situating them within the framework of the national curricula and uses of geographers in France.

To achieve this, we analyze a wide range of scholarly and institutional texts published since the seventeenth century. Scholars interested in education, producers of methods, and polygraph authors sensitive to geography have thought about various methods for entering into knowledge of the world. From the nineteenth century, scientific exchanges around geography pedagogies unfold, particularly on the occasion of exhibitions or scientific conferences. Our bibliographic choices take into

J.-P. Chevalier · G. Labinal (✉)
EMA, CY Cergy Paris Université, Cergy, France
e-mail: jean-pierre.chevalier@cyu.fr; guilhem.labinal@cyu.fr

M. Hardouin, C. Leininger-Frézal (eds.), *Curricular Changes in Geography Education*, International Perspectives on Geographical Education,
https://doi.org/10.1007/978-3-031-96551-7_7

account this diversity of expression frameworks and consider the epistemological particularities of geographical education in France, which are influenced by the double rupture in secondary education: between life sciences and Earth sciences, and between disciplines that take into account emotions, particularly those of artistic education.

7.2 Visual Perception Is Prioritized in Geographic Learning

The practices in school geography and the definition of geographical knowledge are epistemologically closely linked. In the seventeenth and eighteenth centuries, before geography became a school subject in Europe and the Americas, sight was the primary sense mobilized. Some other senses were mobilized when geography became a school subject, to support memorization work.

7.2.1 Observations on the Map

In the seventeenth and eighteenth centuries, geography was presented as "a very proper study for children, because it depends a lot on the senses, and we make them see through the eyes" (we traduce) (Nicole, 1670). This reference to the senses refers to the sight of maps which, with the eyes, are enough to learn (Martineau Du Plessis, 1700). Seeing and showing on the map is, in fact, "the only way to fix the imagination of children" (Lenglet Du Fresnoy, 1736). In this context, the didactic device proposed by Bon de Bouis (1737) appears singular because it calls upon displacements. Bon de Bouis features boxes, ribbons, and pyramids in the Tuileries Garden in Paris: each of these objects, supposed to represent a country or a city, is placed in the garden following the topographical arrangement of what it symbolizes. However, he does not mention in his text the kinesthetic memory of the wanderings, which differentiates his device from the virtual movements made elsewhere on the maps.

7.2.2 Rhyming Artificial Verses for Teaching Geography

Pedagogues faced with the encyclopedic image of geography they seek to teach have multiplied mnemonic devices. One such device is the artificial verse, which relies on the sonic quality of rhymes to memorize geographical inventories. In French, Claude Buffier (1729) employed this method as early as the beginning of the century. James Bisset also employs phonetic resonances in his rhymes to author his *Geographical Guide* (1805) in the English language. He structures his argument through an imaginary journey, and although the book contains few illustrations, the author of images suggests to the reader that they will be seeing things through the verbs he uses.

> A New geographical method I'll try,
> Which, on a slight glance, will at once give a notion
> What maritime cities and ports you'll espy,
> As fam'd British Islands you skirt on the ocean
> I therefore propose to engage a snug ship—
> My readers embark on a party of pleasure;
> Round England and Scotland then send them a trip,
> When ev'ry fam"d Port they may view at their leisure
> (James Bisset, 1805).

The artificial verses process (Flèche, 1838; Parley, 1845, lesson IX) is gradually being abandoned. If, in his book, James Bisset invites his reader to "see," the engravings are rare, as in most books from the beginning of the nineteenth century (Meissas & Michelot, 1827).

7.2.3 *Observing Geographic Images*

Pictures are found more in books for younger students where they hold an important place. In his *Geography for Beginners,* Peter Parley wrote in 1845: "Before we proceed, please look at a few pictures." During this era when geography curricula were still imprecise, methodical geography authors drew inspiration from one another, as illustrated not only by the text of Parley but also by that of Spiess and Cros (1849). The text in French follows the same approach: "Carefully examine the small engravings" (we traduce) (see Fig. 7.1).

This approach is based on an intuitive pedagogy, referred to as "things-based lessons" or *anschauungspädagogik* in German, which emphasizes experiential learning through direct observation and hands-on activities. On both sides of the

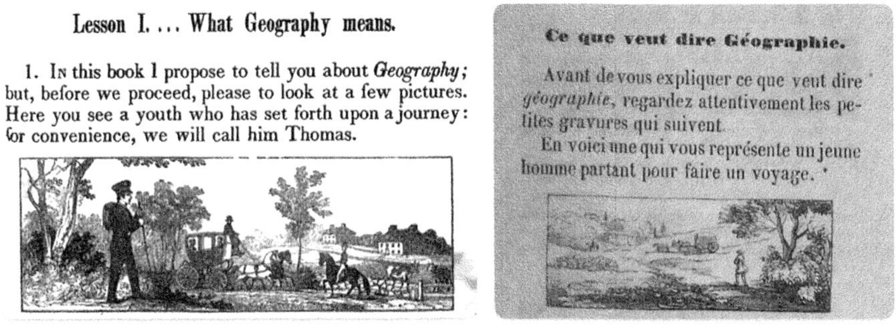

Lesson I. . . . What Geography means.

1. In this book I propose to tell you about *Geography;* but, before we proceed, please to look at a few pictures. Here you see a youth who has set forth upon a journey: for convenience, we will call him Thomas.

Ce que veut dire Géographie.

Avant de vous expliquer ce que veut dire *géographie,* regardez attentivement les petites gravures qui suivent.

En voici une qui vous représente un jeune homme partant pour faire un voyage.

Fig. 7.1 Partley 1845, *Geography for Beginners*. New York: Huntington and Savage, p. 7 (left); Spiess and Cros (1849), *Elements de géographie*. Paris: Crumer, p. 3 (right). In France and the United States, geography textbook authors introduce beginners to images. This approach makes visual perception the primary entry point for discovering the world

Atlantic, text atlases are published, organized on the double-page mode (Swinton, 1875), with maps and engravings placed next to the text. In France, this model is also becoming widespread in primary education (Foncin, 1875). Advances in printing contribute to the dissemination of images of all kinds. In 1882, the French primary inspector B. Berger (1982) underlined the advance of German publishers for the quality of their wall pictures. But it was only at the end of the nineteenth century that the term *landscape* appeared in geography, as well as the iconography of typical landscapes.

7.2.4 Associating Gaze and Gesture

Cartographic exercises contribute to the learning of geography (Lamotte, 1832). Alexandre Le Béalle (1851) emphasizes the combination of gesture and gaze. The sketch, drawn on the blackboard during the lesson, is a canonical practice. Émile Levasseur demonstrated this during the Paris Universal Exposition in 1878 (Levasseur, 1878). The sketch must primarily be reproduced by the student to reinforce its memorization through the gesture that outlines its trace. Therefore, the practice of tracing without accompaniment by the thought of the geographic device is prohibited (Buisson, 1878–1887). Other methods employ puzzles to complete blank maps (Oberlin, circa 1790). Some prioritize the memorization of shapes from a geometric template (Apgar & Apgar, 1866). Through the manipulation of puzzle pieces and tracings, it is a matter of combining visual memory and gesture.

7.3 Active Pedagogies and the Teaching of Geography

Initiating active observation-based pedagogical approaches in school geography, particularly among younger students, effectuates an epistemological transformation, enabling the incorporation of diverse sensorial perceptions that are often neglected in academic geography.

7.3.1 First the Local Environment and the Scale Models

With the rise of school publishing, many educators reference Rousseau or Pestalozzi (Huber, 1997) who advocate for a holistic and intuitive approach to nature. In his educational novel Émile, Jean-Jacques Rousseau criticized the teaching of geography from maps. He proposes investigating the immediate environment as the starting point for geography instruction, but multisensory approaches are absent from this novel. Following his lead, many educators suggest beginning geography studies with an examination of the immediate environment (Darroch, 1906; Capel, 1981), yet they focus primarily on the visible. Pestalozzi has his pupils create clay reliefs after a stroll as a step toward map reading (Vulliemin, 1871) This practice of relief-making is continued with sand-filled boxes (Clozier, 1937), relief modeling (George, 1938), or cardboard cutouts (Diville, 1938). Through this process, combining movement and thought, various types of models are created, such as a classroom, an imaginary country, or a typical landscape (Bonnesoeur, 1899).

7.3.2 Walking, Listening, and Feeling

An important inflection is brought by Rodolphe Töpffer who offers his students to "see the world, get some fresh air, stroll, digest what we learn, observe, link science to life" (we traduce) (Töpffer, 1844). In the accounts of the excursions of the troop of his boarding school, Rodolphe Töpffer evokes what one smells, feels, or hears. He tells us about multiple sensory perceptions: the fatigue of long walks, cold or hot, wet and dry, the sounds of rain and glaciers, the ringing of bells, and music. Töpffer also reports some smells, of incense or garlic. The perceptions he relates are not limited to sight. Töpffer's school caravans are then widely presented by Ferdinand Buisson, director of primary education in France (Buisson, 1878–1887). *Promenades (walks,* we traduce) (Chaumeil, 1880–1887) and *Voyages scolaires* (*school trips,* we traduce) (Durand, 1880–1887) are also entitled to their article in Ferdinand Buisson's dictionary. Edmond Blanguernon, then at the head of primary education in the rural department in Champagne (France), enrolled in the school timetable twice a month (Blanguernon, 1913). The "*classe promenade*" participates in the education of the senses, sight, taste, smell, and hearing, associated with school subjects including geography.

In German or French-speaking countries (Anker, 1872; Freinet, 1994), school promenades have spread somewhat in primary education (see Fig. 7.2). The presence of a single teacher promotes the decompartmentalization of school disciplines and teaching practices opening the school to its environment. Lucien Gachon (1927), who transitioned from a primary school teacher to a university professor of geography, pioneered sensory exploration classes in the 1920s (Zaremba, 2002). These classes entailed experiential learning exercises where he and his students attentively observed and documented the multisensory qualities of their environment, including the olfactory, auditory, and acoustic aspects.

In secondary education, the practice of geography field trips led by the geography teacher remains relatively uncommon, despite the encouragement from French Minister Jean Zay's circulars (George, 1938). Conversely, preschool education devotes a systematic focus to sensory education (Maucourant & Jolly, 1920).

Fig. 7.2 *Der Schulspaziergang*, A.S. Anker (1872). Museum Oskar Reinhart, Winterthur, Switzerland. (Reproduced from Creative Commons file. Von Anker bis Zünd, Die Kunst im jungen Bundesstaat 1848–1900, Kunsthaus Zürich)

7.4 Toward a Multisensory Geography

Since the 1960s, the Commission for Geography Education of the International Geographical Union has promoted approaches based on direct observation, as "observing what is under one's nose" does not take into account auditory, olfactory, or thermal perception-related information (Graves, 1966). While the direct observation of certain phenomena may be perceived as a "concrete basis" that should not be overlooked (Casanova, 1964), pedagogical practices that explore learning modalities integrating gesture, listening, and walking have not been disseminated widely in geography education.

7.4.1 Geography Learning Remains Fundamentally Dependent on Visual Processing

Following the French educational reform of the "third pedagogical time" in 1968,[1] activities aimed at awakening elementary school students' senses developed, but they were insufficient to promote multisensory approaches. A seminal work, published at the beginning of the 1970s, suggests favoring a "wake-up call" for the five senses to test their limits (Porcher et al., 1971, p. 169). However, pedagogical tools

[1] Olivier Guichard, ministerial decree of August 7, 1969.

used to study the environment and the surroundings rarely implement this approach. As George noted, "the history of geography is a long history of images" (we traduce).[2] Even when conducted within the framework of a pedagogical outing, landscape study primarily relies on visual identification and characterization of geographic entities. Until the end of the 1990s, non-visual perceptions were mostly excluded from usual pedagogical practices.

7.4.2 A Geography of Sonic, Spatial, and Olfactory Landscapes

When geographer Yi-Fu Tuan (1974) coined the concept of Topophilia to study the environment, focusing on the affective bonds between humans and their surroundings, the 1970s saw the emergence of a perceived and lived geography (Metton, 1974; Frémont, 1976). However, after André Siegfried's *Géographie des couleurs et des sons* (*Geography of Colors and Sounds*, we traduce) (1950), geographers offered limited theoretical contributions that could be applied to teaching. In France, it was late in the day before they expanded the scope of perceptions by examining odors in a pioneering work (Duleau & Pitte, eds., 1998). Jean-François Staszak (1998) noted that there is a "kind of relationship between vision and space, and that geographers work their eyes more than their other senses" (cited by Di Meo, 2000, p. 331, we traduce). This focal point partially explains why early geographers working on this type of issue found themselves isolated, as observed by Kay Anderson and Susan J. Smith (2001) when they questioned the possibility of developing an emotionally sensitive geography program in Anglophone geography.

Two decades later, existential geography, as conceptualized by Dardel (1952), began to take shape, with a focus on lived experience (Besse, 2003, pp. 7–8). Long neglected despite initial works by composer and pedagogue Raymond Murray Schafer (1977), who invented the concept of "Soundscape" (echoing "Landscape"), geographers started examining music (Guiu, 2006), sounds, and sonic ambiances (Augoyard, 1991; Roulier, 1999; Masson, 2007, 2009; Jolivet, 2010; Law, 2013). Without necessarily having to abandon visual approaches (Phillips, 2015), they evoked "Sensuous Geographies" (Rodaway, 1994), attempted to grasp the olfactory sensoriality of landscapes (Henshaw, 2013; Fraigneau, 2019), and experimented with multi-sensory approaches to urban space experiences (e.g., Grésillon, 2005, 2012; Manola, 2013). Parallel to the growth of Anglophone research on emotions (notably Davidson et al., 2007; Davidson & Milligan, 2004), these studies contribute to the questioning of artistic and sensitive cartography, integrating wandering into certain research devices (Olmedo, 2015). These issues were slow to be tackled

---

[2]George, P. (1989). *Les hommes sur la terre, la géographie en mouvement*, p. 202. Cited by Giolitto, P. (1992). *Enseigner la géographie à l'école*. Paris: Hachette.

directly in geography (Guinard & Tratnjek, 2016), but sensory-oriented approaches have developed in experimental pedagogical settings within geography.

7.4.3 Aligning Educational Initiatives with Advances in Scientific Geography

In France, the landscape approach developed by geographer Charles Avocat (1982, 1983) already provided key elements for educators seeking to integrate senses beyond vision. By establishing a methodology for approaching the landscape (what one sees and, secondarily, what one hears or feels), he effectively distinguished between a "*paysage réel*" (Real Landscape, we traduce), an objective and universal concept, and a "*paysage image*" (Landscape Image, we traduce), which is more personal (Avocat, 1983, pp. 14–15). By placing perception at the center of landscape analysis, Charles Avocat proposed an approach that integrated subjective and aesthetic variables related to color appreciation, light, movement, texture (natural materials), as well as sounds and smells to define a landscape ambiance. Although deceased prematurely, he was unable to further develop his work, but his multisensory observation grid was disseminated. Adaptations of this framework have been implemented in training and education.

Except for the use of models as learning tools in geography (Mercier, 1992) or school outings that involve visual analysis of the landscape, multisensory initiatives have remained isolated for a long time. However, within the context of a renewal of pedagogical approaches and the reevaluation of traditional learning spaces, research has been conducted on artistic practice in schools (Gaujal, 2016) and sensitive outings. In his doctoral dissertation, Médéric Briand (2016, pp. 78–79) observes that a middle school child learns through embodied experience, noting that "fatigue in pedestrian experience" allows them to perceive relief through proprioception while walking uphill, downhill, or on level ground (*Ibid.*). Therefore, inspired by Rachel Thomas's work (1999, 2010) on architectural and urban ambiances, Briand blindfolded students during outings to force them to rely on "other senses such as hearing, touch, and smell" (we translate) (Briand, 2016, p. 342). He also uses a scent journal that students insert into a sensory notebook to describe their tactile perceptions. In a similar vein, Julie Picard (2019, p. 3) has explored the value of the field trip as an experimental and pedagogical device, using drawing to record students' experiences during a walking itinerary in Bordeaux. In teacher training institutes, geography through wandering and perception is being implemented to account for lived landscapes (see Fig. 7.3).

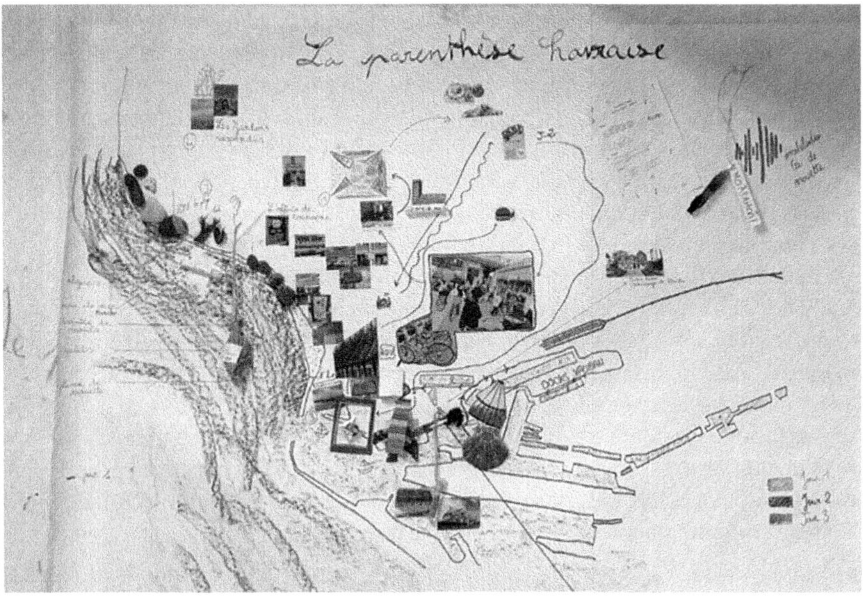

Fig. 7.3 Sensitive Map of the City of Le Havre. Map produced by trainee teachers during a trip organized by Anne Rebouillat (history-geography teacher trainer) and Frank Lafon (visual arts teacher trainer). National Higher Institute of Teaching and Education of the Academy of Versailles, CYU Cergy Paris University. Photograph taken on April 29, 2021 (G. Labinal). Sound perception is integrated into the production (the cries of seagulls), as well as touch (seaweed, pebbles, feathers, etc.) or wandering as part of an itinerary

7.5 Discussion

The analysis of past practices demonstrates that the use of senses other than sight in the construction of geographic knowledge has been present for a long time. However, the limited mobilization of different senses in teaching has been reinforced by the diversity of epistemological traditions within the field of geographical education, which has, sometimes, been a compulsory subject and sometimes an optional one. This has led to a vast disciplinary field that has been oriented toward social sciences or environmental studies (Chevalier, 2000). The separation in France between the teaching of arts and that of geography can explain why we do not find any trace of use in geography of a musical work such as *Entrance to the Harbour*[3] or other recordings from the *World Soundscape Project* (Truax et al., 2004). Jean-François Augoyard (1991) highlighted the primacy of sight over hearing, which informs us about the direction of a source but more difficulty about its distance.

[3] The *Vancouver Soundscape anthology* was edited by Murray Schafer in 1973 as part of the initiative developed within the educational and research group he led at Simon Fraser University (World Soundscape Project).

While visual aids have long been available for reproducing images of the world, procedures for recording and using sounds are relatively recent. The other senses (smell, temperature, humidity, etc.) did not benefit from techniques for reproduction that could be used in the classroom, often limiting sensory experiences to the local space of students. Only outdoor classes allowed for characterizing places or directions by analyzing sounds and smells. Touch and thermoreception (the perception of heat and cold by the skin) play a notable role in the orientation of the blind and those with impaired vision.

To account for this epistemological diversity within geographical education, our study focused on the French curriculum from the twentieth century onward, where multi-sensory approaches began to appear or resurface in practices while remaining marginal. Although taste, pain, or balance perception do not intervene much in geography teaching, intuitive methods have started to mobilize other perceptual senses to study students' environments. However, the official geography curriculum for students aged 9–12 years old still does not mention multi-sensory approaches, as if they were not compatible with traditional school forms (Vincent, 1980). For older students, the situation is expected to evolve rapidly.

7.6 Conclusion

An examination of theoretical proposals, educational materials, and pedagogical practices from previous eras reveals that, while the teaching and learning of geography may have predominantly emphasized visual approaches, other methods have been developed or can be used to engage multiple senses. However, students were rarely encouraged to distinguish the contributions of the five senses in characterizing landscape ambiances by rationalizing the articulation, concordance, or potential disparities between perceived elements during field trips. Despite these legacies, theoretical and practical tools are now being designed in the field of geography education to promote experiential approaches (Leininger-Frézal, 2019). The current proliferation of experiential-based announcements on professional platforms undoubtedly signals a possible evolution in the complexity of the approach proposed for grasping inhabited spaces through sensory means. Beyond our historiographic study, it is essential to investigate and harness these contributions, share them, and deploy them to their full potential.

References

Anderson, K., & Smith, S. (2001). Editorial: Emotional geographies. *Transactions of the Institute of British Geographers, 26*(1), 7–10.

Anker, A. S. (1872). *Der Schulspaziergang*. Museum Oskar Reinhart.

Apgar, E. A., & Apgar, A. C. (1866). *Improvment in geographical maps*. United States patent. US58648.
Augoyard, J. F. (1991). La vue est-elle souveraine dans l'esthétique paysagère. *La théorie du paysage en France (1974-1984)*. (Roger, 1st ed., 334–345). Champ Vallon.
Avocat, C. (1982). Approche du paysage. *Revue de Géographie de Lyon, 57*(4), 332–347.
Avocat, C. (1983). Essai de mise au point d'une méthode d'étude des paysages. *Actes du colloque Lire le paysage, Lire les paysages, 24 et 25 novembre 1983 Université de Saint-Etienne*, 11–31. CIEREC.
Berger, B. (1982). *Tableaux muraux d'enseignement*. (Dictionnaire de Pédagogie et d'instruction primaire part1(2), 2850–2851. Buisson, Ed.). Hachette.
Besse, J. M. (2003). *Face au monde : Atlas, jardins et géoramas*. Desclée de Brouwer.
Bisset, J. (1805). *Geographical guide: A poetical, nautical trip round the Island of Great Britain*. J. Harris.
Blanguernon, E. (1913). *Pour l'école vivante*. Hachette.
Bon de Bouis. (1737). *Le Parterre géographique et historique, ou Géographie-pratique, ou Nouvelle méthode d'enseigner la géographie & l'histoire*. Nyon fils.
Bonnesoeur, A. (1899). *Photographie du relief construit dans l'école communale de Créance en 1882 par le directeur M. Le Petit*. https://gallica.bnf.fr/ark:/12148/btv1b85538167.item
Briand, M. (2016). *Approche sensible en sortie. Construire une relation au terrain en géographie scolaire à l'école élémentaire*. Acteurs et action Perspectives en didactique de l'histoire et de la géographie. (Thémines, J.-F. & Doussot S. Ed.), 73–96. Presses universitaires de Caen.
Buffier, C. (1729). *Géographie universele Exposée dans les différentes Métodes qui peuvent abréger l'étude & faciliter l'usage de cette Sience* (4th ed.). Giffert.
Buisson, F. (1878–1887). Dictionnaire de pédagogie et d'instruction primaire 1st part. 138. (Buisson, F. Ed.). Hachette.
Capel, H. (1981). *Institutionalization of geography and strategies of change*. (Geography, ideology and social concern, 37–69. Ross Stoddart, Ed.). Basil Blackwell.
Casanova, A. (1964). *La géographie en 6ᵉ et les techniques de connaissance*. (L'enseignement dans le cycle d'observation. Rouchette & Vacquier. Ed.). *Cahier de Pédagogie moderne, (27)* 109–113. Bourrelier, A. Colin.
Chaumeil, J. (1880–1887). *Promenades*. (*Dictionnaire de pédagogie et d'instruction primaire* 1st part. 2457–2458. Buisson, Ed.). Hachette.
Chevalier, J.-P. (2000). La géographie dans les programmes scolaires en Europe. *Cybergeo, European Journal of Geography* (130). https://journals.openedition.org/cybergeo/213
Clozier, R. (1937). Exercices pratiques. 1. La caisse de sable. *L'Information géographique, 2*(1), 29.
Dardel, E. (1952). *L'homme et la terre* (1st ed.). Presses Universitaires de France.
Darroch, A. (1906). The teaching of geography. *The Scottish Geographical Magazine, XXII*(9), 484–490.
Davidson, J., Bondi, L., & Smith, M. (Eds.). (2007). *Emotional geographies*. Ashgate. 258 p.
Davidson, J., & Milligan, C. (2004). Embodying emotion, sensing space: Introducing emotional geographies. *Social and Cultural Geography, 5*(4), 523–532. https://doi.org/10.1080/1464936042000317677
Di Meo, G. (2000). Comptes rendus. *Annales de géographie, 109*(613), 331.
Diville, W. (1938). *Les exercices de modelage et de cartonnage*. (L'enseignement de la géographie. Sorre. Ed.), 63–68. Bourrelier.
Duleau, R., & Pitte, J.-R. (1998). *Géographie des odeurs*. L'Harmattan.
Durand, H. (1880–1887). Voyages scolaires. (*Dictionnaire de pédagogie et d'instruction primaire 1st part*, 2985–2986. Buisson, Ed.). Hachette.
Flèche, P.-J. (1838). *Petite géographie méthodique de la France, en vers artificiels, comprenant les quatre-vingt-six départements sur 86 rimes différentes, avec des notes explicatives... et les possessions françaises hors d'Europe aussi en vers accompagnés de notes*. Lecoffre.
Foncin, P. (1875). *La Première année de géographie*. Colin.

Fraigneau, V. (2019). La sensorialité olfactive du paysage, médiatrice d'une reliance sensible. *VertigO, 19*(3) http://journals.openedition.org/vertigo/27072
Freinet, C. (1994). *œuvres pédagogiques 1*. Seuil.
Fremont, A. (1976). *La région espace vécu*. Presses Universitaires de France.
Gachon, L. (1927). En classe promenade. *Journal des Instituteurs et institutrices, 28*, 402–403.
Gaujal, S. (2016). *Une géographie à l'école par la pratique artistique*. Doctoral thesis, Université Sorbonne Paris Cité.
George, P. (1938). La place de la géographie dans les loisirs dirigés. *L'Information géographique, 4*, 171–173.
Graves, N. J. (1966). *Les méthodes d'enseignement : l'observation directe. L'enseignement de la géographie,* 44–79. (Commission de l'enseignement de la géographie de l'UGI. Ed.). Unesco-Ipam.
Grésillon, L. (2005). *Sentir Paris : bien-être et matérialité des lieux*. Doctoral thesis, Université Paris 1.
Grésillon, L (2012). *Sentir son lieu (ou ne pas le sentir). De l'importance de la sensorialité.* (La Fabrique des modes d'habiter, 91–105. Morel-Brochet & Ortar. Ed.). L'Harmattan.
Guinard, P., & Tratnjek, B. (2016). Géographies, géographes et émotions. Retour sur une amnésie... passagère ? *Carnets de Géographes* (9). https://journals.openedition.org/cdg/605
Guiu, C. (2006). Géographies et musiques Quelles perspectives ? *Géographie et culture, 59*.
Henshaw, V. (2013). *Urban smellscapes: Understanding and designing city smell environments*. Routledge.
Huber, B. (1997). Une étape déterminante dans l'évolution de l'enseignement de la géographie : J.-H. Pestalozzi à Yverdon (1805-1825). *Geographica Helvetica, 4*, 129–132.
Jolivet, V. (2010). *Miami la cubaine ? : pouvoir et circulation dans une ville carrefour entre les Amériques*. Doctoral thesis, Université Paris 1.
Lamotte, M.-L. (1832). *La géographie enseignée par le dessin ; ou méthode facile et sure d'apprendre la géographie par la copie de cartes graduées et disposées pour cet enseignement*. Hachette.
Law, L. (2013). *Urban senses*. (Introducing Human Geographies. 706–719, Cloke et al. Eds.) Routledge.
Le Béalle, A. (1851). *Cours de dessin classique élémentaire des cartes géographiques*. Delalain.
Leininger-Frézal, C. (2019). *Apprendre la géographie par l'expérience : la géographie expérientielle*. Ability to conduct researches. Université de Caen France.
Lenglet Du Fresnoy, N. (1736). *Géographie des enfans ou méthode abrégée de la géographie*. Rollin.
Levasseur, É. (1878). *L'enseignement de la géographie dans l'école primaire. Conférences faites à la Sorbonne aux instituteurs délégués à l'exposition universelle de 1878*. Delagrave.
Manola, T. (2013). Conditions et apports du paysage multisensoriel pour une approche sensible de l'urbain. *Carnets de géographes5*. http://journals.openedition.org/cdg/1107
Martineau Du Plessis, D. (1700). *Nouvelle géographie ou description exacte de l'univers....* Gallet.
Masson, D. (2007). *Musiciens en mouvement. Pratiques sonores en métro.* (Musiques Patrimoine des expériences culturelles urbaines, Cheyronnaud & Pecqueux Eds.) Ecole nationale d'architecture. https://shs.hal.science/halshs-00441991/document
Masson, D. (2009). *La perception embarquée. Analyse sensible des voyages urbains. Sciences de l'Homme et Société*. Doctoral thesis, Université Pierre Mendès-France. https://theses.hal.science/tel-00441955/file/Masson_These.pdf
Maucourant, M.-B., & Jolly. (1920). *L'éducation des sens par l'activité. Pratique. Exercices. Résultats*. Fernand Nathan.
Meissas, A., & Michelot, A. (1827). *Nouvelle géographie méthodique destinée à l'enseignement....* Hachette.
Mercier, C. (1992). *La maquette comme instrument d'apprentissage. Un géographe relit Piaget.* Enseigner l'histoire et la géographie. Un métier en constante rénovation Mélanges offerts à Victor et Lucile Marbeau. 121–125. Audigier et al. (eds.). A.F.D.G.

Metton, A. (1974). L'espace perçu : diversité des approches. *L'Espace Géographique, 3*, 228–230.
Murray Schafer, R. (1973). *Entrance To The Harbour (The Vancouver Soundscape)*, two records, Sonic Research Studio, Simon Fraser University, Burnaby 2, B.C., Canada.
Murray Schafer, R. (1977). *The tuning of the world (the soundscape)*. Knopf.
Nicole, P. (1670). *De l'éducation d'un prince : divisée en trois parties, dont la dernière contient divers traittez utiles à tout le monde*. Veuve Charles Savreux.
Oberlin, J-F. (1790 circa). Fond de carte vierge de la France. Archives de Strasbourg 77 Z 99/20.
Olmedo, É. (2015). *Cartographie sensible. Tracer une géographie du vécu par la recherche-création*. Doctorat thesis, Université Paris 1-Panthéon Sorbonne.
Parley, P. (1845). *Parley's Geography for beginners...* Huntington and Savage. https://digital.library.pitt.edu/islandora/object/pitt%3A00z357449m/viewer#page/8/mode/2up
Picard, J. (2019). (Faire) enseigner la géographie hors-les-murs : initier à la sortie sensible en master MEEF 1er degré. *Feuilles de géographie, 6*, 17 p.
Phillips, R. (2015). Playful and multi-sensory fieldwork: seeing, hearing and touching. *New York Journal of Geography in Higher Education, 39*(4), 617–629.
Porcher, L., et al. (1971). *Pour comprendre et pratiquer les activités d'éveil*. Armand Colin.
Rodaway, P. (1994). *Sensuous geographies: Body, sense and place*. Routledge.
Roulier, F. (1999). Pour une géographie des milieux sonores. *Cybergeo: European Journal of Geography* [En ligne], Environnement, Nature, Paysage, document 71, mis en ligne le 21 janvier 1999, consulté le 05 juillet 2024. http://journals.openedition.org/cybergeo/5034
Siegfried, A. (1950). Géographie des couleurs et des sons. *La Revue de Paris, 2*(6), 3–11.
Spiess, C., & Cros, A. (1849). *Éléments de géographie*. Curmer.
Staszak, J.-F. (1998). Pistes pour une géographie des odeurs. (Géographie des odeurs, Duleau & Pitte Eds.). L'Harmattan, 49–58.
Swinton, W. (1875). *A complete course in geography, physical, industrial, and political with a special geography for each state*. Ivison, Blakeman and company.
Thomas, R. (1999). Cheminer l'espace en aveugle : corps stigmatisé, corps compétent. *Alinéa, 9*, 9–24.
Thomas, R. (2010). *Marcher en ville. Faire corps, prendre corps, donner corps aux ambiances urbaines*. Edition des archives contemporaines.
Töpffer, R. (1844). *Premiers voyages en zigzag ou Excursions d'un pensionnat en vacances dans les cantons suisses et sur le revers italien des Alpes*. Dubochet et Cie.
Truax, B. et al. (2004). World soundscape project. *The Canadian Encyclopedia*. https://www.thecanadianencyclopedia.ca/en/article/world-soundscape-project
Tuan, Y.-F. (1974). *Topophilia. A study of environmental perceptions, attitudes and values*. Prentice-Hall.
Vincent, G. (1980). L'école primaire française, Étude sociologique. Presses universitaires de Lyon, Éditions de la Maison des Sciences de l'homme, 344 p. (Ouvrage épuisé. Édition numérique 2006 : http://pressesuniv-lyon2.fr).
Vulliemin, L. (1871). *Souvenirs racontés à ses petits-enfants*. Bridel. https://books.google.fr/books?id=ER1fIkovTRIC&printsec=frontcover&hl=fr&source=gbs_ge_summary_r&cad=0#v=onepage&q=g%C3%A9ographie&f=false
Zaremba, J-L. (2002). Lucien Gachon (1894-1984) instituteur, précurseur de la classe promenade, défenseur de l'école rurale. Doctoral thesis, Université Lyon 2.

Jean-Pierre Chevalier is professor emeritus at CYU Cergy Paris University (France). Member of the ÉMA laboratory (École, Mutations, Apprentissages). His research mainly focuses on geography teaching and the history of geography teaching. He was director of the teacher training institute of the Versailles Academy. He participates in cooperation missions in Africa in the field of educational research.

Guilhem Labinal is an professor in geography at CYU Cergy Paris Université (France). He previously served as the head of the history and geography department at the INSPÉ de l'académie de Versailles (CYU) and co-director of the ÉMA laboratory (École, Mutations, Apprentissages). His research has focused on the history, epistemology, and pedagogy of geography, and he is currently particularly interested in learning environments.

Chapter 8
Systemic Approach of Globalization Through Worldwide Landscapes Analysis

Elisabeth Dury

8.1 Introduction

Defining the societal challenges due to globalization and teaching students how to identify their markers is now what is expected in geography education. The study of landscapes has always been a key element in both academic geography and school geography, as a means of tackling ever-changing issues. It moved beyond basic landscape descriptions to integrate successively quantitative and modelling approaches in the 1960s, then cultural turn factors in the 1980s, or even systemic analysis. The plasticity of the landscape as a framework for analysis suggests that it should find a legitimate place in explaining contemporary challenges of globalization. To achieve this, landscape analysis must meet three requirements: First, comparisons should be made between different situations around the world, taking advantage of the technological innovation of the Internet; second, the focus should be on points of contact and exchange as key interface landscapes of globalization; and third, the most should be made of the infatuation with landscape which is visible in the mobilization of citizens at local level against projects deemed to be damaging to the landscape (e.g., wind farms) or, conversely, for projects enshrining landscapes (e.g., national competitions such as "favourite village of French people").

Coastal landscapes seem to be an effective point of entry in order to combine the construction of knowledge, didactic renewal, and civic aims (Heimberg, 2011) of landscape analysis. As archetypes of the globalization interface, the economic value of coastal landscapes continues to increase with successive integrations, while the world's population continues to move inexorably toward the coast, exposing itself even more to the risks associated with climate change in particular. The resulting

E. Dury (✉)
EA 2076 HABITER; Université de Reims Champagne-Ardenne, Reims, France
e-mail: elisabeth.dury@ac-reims.fr

M. Hardouin, C. Leininger-Frézal (eds.), *Curricular Changes in Geography Education*, International Perspectives on Geographical Education,
https://doi.org/10.1007/978-3-031-96551-7_8

conflicts shape these landscapes while also causing citizens to mobilize. Images from the many webcams scattered along the world's coastlines are a mine of freely accessible information and constitute perfect tools that can be used as part of a teaching sequence. The aim of this chapter is, therefore, to show the didactic relevance of landscape analysis using webcams to decipher the societal challenges posed by globalization, based on a selection of examples mainly from coastlines.

First, the focus will be on the particularities of landscapes captured by webcams, highlighting the limitations of geographical analysis. Then, based on classroom work, this chapter will show which globalization issues students may come up with after comparing selected coastal landscapes (integration, ranking process, threats) and how this can be a systemic tool in mobilizing and building knowledge. Finally, we will see that this type of remote landscape analysis can have an impact on students' civic engagement and be a powerful lever for mobilizing cross-curricular skills (critical thinking, media literacy) to help train enlightened, active citizens capable of defending societal choices.

8.2 A Landscape with Many Variables

8.2.1 A Major or Minor Departure from Geographers' Definition of Landscape?

The discussion is based on a definition of landscape as an "arrangement of visible objects perceived by a subject" (Brunet et al., 2005), a "datable cultural construction" (Berque, 1990), and a "material arrangement of natural and social space" (Lévy & Lussault, 2003). In a nutshell, the landscape is "a central element in a cultural system, … it acts as a signifying system through which social systems communicated, reproduced, experienced and explored" (Lewis & Duncan, 1992). This mediating relationship between the materiality of the landscape and a system of symbols, representations, and issues means that the landscape can become a tool for reading geography and provide access to the meaning of dwelling in any type of landscape, from the most ordinary to the most remarkable.

It is only a short step from there—and a long-standing ambition—to consider landscape as a vector toward an active civic role. It was already apparent in the institutional prescriptions for education to sustainable development, which early on seemed to favor a tool revealing the ways in which territories are dwelled (Biaggi, 2015), combining practices, representations, and actors in constant interaction. In France, the rewriting of secondary school curricula around the topic of dwelling has reinforced this, making landscape studies a means of understanding the conflicts between actors in relation to potential or actual developments (Sgard, 2015).

However, the proliferation of zenithal or oblique views and satellite images at school considerably limits our understanding of the observer's position: There is a profound gap between the desired aim and the method used (Biaggi, 2015). If we plan to renew the value of landscape studies as a tool for revealing the way people are dwelling, then we must also argue for refocusing the gaze from the observer as a horizontal relay of the eye. Through photographs of La Baule, O. Lazzarotti (2014) reminds us that the shots taken are primarily the work of local people who experience a place and write a discourse on the landscape in images, deliberately favoring one framing rather than another, which undeniably argues in favor of a return to the land or the search for alternatives to aerial or satellite views.

The use of geographical information technologies as geo-technologies (Genevois, 2020) in geography education is a case in point. The use of tools for geographical exploration (virtual globes) or localization Global Positioning System (GPS) has become commonplace in the classroom; the use of webcams is probably less so. The few webcam exercises for a classroom usually involve repeated observation over a long period of time, for example, seasons (Sawyer et al., 2010).

In contrast, the social use of webcams in the private sphere, for example, to finalize the choice of a holiday destination, or in the professional sphere (e.g., promotion of a property sale), is trivial, even unmarked. Using a free access webcam is sometimes part of everyday practice for users, and the popularity of this medium is a reality that needs to be taken into account. Its transfer to the field of producing geographical knowledge of the living environment through landscape studies requires that certain points be clarified.

The first concerns the relevance of webcams for landscape studies, which trivially amounts to asking whether a webcam can really be used to understand a landscape. The real sticking point is the comparative advantages of still and moving images. In this respect, the questioning is identical in all points to that of the legitimacy of video media in the 1980s. Traditionally, a landscape is studied on the basis of a limited number of shots. With a webcam, the number of shots taken depends first and foremost on the image renewal interval, which varies enormously from one site to another and from one operator to another, ranging from a few minutes apart to several times a day. In any case, the idea of the observer entering the landscape *in situ* and at a distance via a webcam is a myth. Livecams are rare. For technical reasons, all of the images are delayed.

But the viewer still has the option of choosing the webcam (sometimes several angles of view are possible for a single location), or panning or zooming in remotely. While choosing a viewpoint, the viewer moves from subject to dweller, actor in his or her own spatial practices. If we start from the principle that the spatialities of an individual determine his or her experience of dwelling in a space, then the material support for the study increases rather than restricts the possibilities (Lussault, 2019). The exploratory dimension fully realized by the use of webcams, however partial, reinforces spatialities. The landscape apprehended by the webcams reveals this.

The second question concerns the objectification of empirical knowledge. There is still the idea that the webcam is a learning artifact, or worse still, a fad. The empirical and present geography resulting from the experience of landscapes through webcams is direct and accessible to the greatest number of people while offering the possibility of comparison between places that are far apart. It stimulates the objectification of empirical knowledge and, in return, creates an environment leading to citizen action. What is more, the transition from a 2D landscape study to a 3D study via webcam reshapes the methods of spatial reasoning and adds value to the cognitive process, particularly in the establishment of chains of logical relationships between the elements observed and identified (Genevois, 2020). In this sense, it is fully in line with the updating of the relationships at the heart of the definition of landscape delineated above.

8.2.2 Taking Advantage of Webcams to Explore a Globalized World Through Tourist Coastlines

There are an estimated 1 billion surveillance cameras in the world. However, not all of them are webcams. A webcam is a digital camera connected to a computer that broadcasts images over a network without any interactivity, as opposed to surveillance cameras, which are only used by public authorities for video surveillance purposes, subject to authorization (Pols, 2018). Scattered across the globe, webcams form a powerful, free network that broadcasts continuously 24 h a day. For instance, the Viewsurf network has been the leading network in French Brittany since 2002, and Visio-environment provides access to a network of 2561 webcams—its own or those of partners—in a wide variety of locations such as beaches, ports, airfields, and so on, with the exception of Africa, a continent that is poorly covered by this service provider. This plethora of technical resources (number of cameras and broadcasters) attracts a large audience: on November 13, 2023, at 3 p.m. French time, 29,679 users were connected to the Worldcam network, which includes 24,231 webcams. Not only are webcam networks the markers of a globalized world, but they give concrete expression to globalization as "a process of generalisation of exchanges between different parts of humanity, between different places on the planet" (Grataloup, 2023). Indeed, it is this acceleration of globalization, characterized by visible flows in a landscape (passers-by, traffic) and informational flows (passing through webcam networks), that has been the most striking feature since the end of the 1990s. These flows, seen as "alternatives to mobility and co-presence," suggest, as J. Lévy (2007) demonstrates, that "the spaces of globalization cannot be reduced ... to those of technology. [They] refer to techniques produced by the actors-users of technologies when they seek to respond to the following problem: how can I take advantage of the World?" Webcams are thus a tool for the connective dimension of dwelling, which D. Boullier (2019) argues that it is consubstantial with dwelling in digital globalization, and which he calls in French the *habitèle*. From this point of view, choosing to work with webcams displaying coastal landscapes is doubly significant: it means taking a gamble on studying interface landscapes that are integrated into and across the world in order to observe how the World dwells these attractive landscapes.

In that respect, it would be a mistake to overlook the wonder generated by the possibility of vicarious travel as part of a learning process: a webcam sweeps us off our feet as much as it sweeps us away, freeing us from distance and borders in a post-COVID era still impacted by the experience of lockdown. Given that only 2.5% of geography sessions take place outside of the classroom, webcams are a credible theoretical and didactic alternative in order to practice geography outside the classroom, if you cannot do it in person (Genevois, 2020).

8.2.3 Challenges and Limits of This Geographical Analysis

However fruitful this approach may be, it does have its limits. In the field, for example, the exploration of dwellings through landscapes can be combined with a multi-sensory approach using sight, hearing, or smell. Through the choices made here, the sensitive exploration of distant landscapes is reduced to a single sense with a webcam: sight. Sound is technically unmanageable. In some cases, the sound of the waves is superimposed on the broadcast images in post-processing, giving the illusion of a soundscape. In reality, we are back to the primacy of the eye in geography which can be regarded as an impoverishment of information.

Moreover, in this respect, there remains an inescapable informational bias that should be mentioned. The law adds a constraint to the technique: with webcams, one may observe not only material elements of the landscape but also the movement and activities of the people present in the frame. Although the added value of landscape surveys using webcams lies precisely in this informative surplus, Article 226-2 of the French Penal Code, for example, sets out rights and obligations regarding the fixation and storage of images. It requires service providers to systematically process images by blurring out faces and silhouettes in order to protect privacy. Strictly speaking, therefore, webcams transmit data rather than information, which can lengthen the time taken to transmit images. Legal conditions may vary from country to country, but on the whole, the working material is already processed by an intermediary.

Finally, it should be borne in mind that webcams are set up to serve local promotion strategies. Broadcasted images are the result of a commercial transaction between a private service provider and the local authorities that commissioned them. As a result, only landscapes deemed worthy of interest by these clients are shown, within the limits of the technical constraints on the deployment of webcams defined by the service provider. This objective necessarily leads to a focus on landscape assets and even to a temptation to overvalue aesthetics. This representation of the beautiful or perfect landscape may hinder the exploitation and objectification of knowledge within the framework of our research or, conversely, be the catalyst for reaction/action on the part of students.

8.3 Guidelines for Bringing Out Globalization Process in Classroom Work

8.3.1 Anchoring in Geography Education and Didactical Beacons

Having laid the foundations for a landscape-based approach to dwelling using webcams, by identifying the relevant epistemological elements in the first section, we now need to look at how they are reflected pedagogically in the curricula. To this end, the *Bulletin Officiel*, the prescriptive document for French secondary education, identifies, among the fields of action of geography teachers, on the one hand, specific skills (*Capacités*) and, on the other hand, transferrable skills (*Compétences*) that belong to the *Socle Commun*, the expected general knowledge basis (*Programme de l'enseignement d'histoire-géographie de la classe de seconde générale et technologique, de la classe de première de la voie générale et de la classe de première de la voie technologique*, 2019).

Concerning specific skills, there are close links to geography first of all, with classic expectations (from finding one's bearings in space to identifying spatial issues), but also to moral and civic education with the aim of mobilizing and developing critical thinking, which is at the heart of learning. Finally, there are links with media and information education (E.M.I.), with emphasis on putting information into context and making informed use of it. Regarding the transferrable skills of the *Socle Commun*, certain objectives linked to the responsible use of digital tools and collaborative spaces are boosted by our approach that revisits the approach to globalization through landscapes.

A final review of the thematic entries in the geography curricula that punctuate the school year and the *Fiches Eduscol* (pedagogic indicative document), however, shows a shift as schooling progresses: The transposition of the aims of geography teaching in the secondary school curricula gradually erases the didactic approach to dwelling in the landscape. Its presence in the possible implementation of the curricula is inversely proportional to that of the concept of globalization. From the very strong choice of a concrete approach in the *6ème* at the *Collège* (Biaggi, 2015) to the possible ups and downs in *4ème* at the *Collège* through the world's urbanization, landscape analysis no longer appears as a possible pedagogical lever by the end of *Collège* and even less so in the *Lycée* (*Programmes d'enseignement de l'école élémentaire et du collège*, 2015).

While globalization forms the notional backdrop to the secondary school curriculum up to the final year of secondary school, there is every reason to believe that an understanding of the factors of differentiation and spatial interaction, along with the societal challenges that it raises, is enough to justify the disjunction with a concrete approach based on dwelling landscapes. It even seems to be paradoxical: the backbone of all secondary school education (*Collège* + *Lycée*) is a set of skills and competences designed to help students become acculturated to the challenges of civic involvement (participation in the *Conseil de Vie lycéenne*, valuing commitment to associations) in conjunction with the gradual increase in the complexity of reasoning (through increasing generality and abstraction), whereas the landscape, a powerful, concrete lever for involvement that is highly valued by society, is sidelined by the institutional mapping of possible educational approaches. To facilitate the didactic transposition of the notion of globalization, I propose to reintroduce landscape analysis, renewed by webcams, at the time of the final years of secondary school (*Lycée*), as a tool for structuring thought and organizing information, making it possible to develop a global vision by shedding light on the experience of spatialities with the aim of constructing knowledge that is both emancipating and mobilizing, that is, participating in the training of the dweller-actor-citizen (Charpentier et al., 2023).

8.3.2 Methods and Setup of the Investigation

The eight coastal webcams selected (Table 8.1) pertain to three categories of places: those that evoke a powerful imagination (Sangatte, Hawaii, Rotterdam, Dubai and Borealis 360), those whose location is assumed to be known to AbiBac students (the island of Sylt on the German-Danish border was the subject of a news item discussed in class), and ordinary places that are most likely unknown to students (St Lunaire, Koh Samui).

The investigation protocol was carried out with 15 volunteer students enrolled in the dual Franco-German AbiBac course at the end of the final year of secondary school, by email and in complete autonomy in terms of duration and management of the various stages. Inspired by S. Joublot Ferré (2021), the protocol was broken

Table 8.1 List of selected webcams

Place/country	Link to the webcam
Lost Bar -Choeng Mon- Koh Samui / Thailand	https://www.worldlivecamera.com/fr/webcam/Choeng-Mon-Koh-Samui.html
Sangatte - Pas de Calais / France	https://www.visionenvironnement.com/livecams/webcam.php?webcam=sangatte
Saint-Lunaire – Pointe du Decolle Ile-et-Vilaine / France	https://www.saint-lunaire.fr/decouvrir/webcam
Port of Rotterdam Nieuwe Waterweg / Netherlands	https://www.youtube.com/watch?v=1sCBeKZjz8I
Hotel Roth am Strand –Westerland-Sytl Schleswig Holstein / Germany	https://www.youtube.com/watch?v=HXnlOrHPAMQ
Borealis 360 – Honningvog University / Norway	https://porthonningsvag.kystnor.no
Mariott International Waikiki Beach – Honolulu Hawaii / USA	https://marriott.ozolio.com/waikiki-beach-marriott/
The Palm –Fairmont / Dubai-United Arab Emirates	https://www.youtube.com/watch?v=7dE4IjDQJmE

down into three stages and required back-and-forth movement between careful observation of the images broadcast by the webcams and the individual restitution of the information in a results table. As a result, it has an immediate reflexive character as well as dynamics of vertical integration (by degressive zooming) and horizontal integration (by contextualizing the reflection) of information.

The three successive phases are as follows (Table 8.2):

1. Before opening the links to the webcams, spontaneous spatial representations of the locations investigated are collected. Students are invited to put into words what they perceive of places before they are confronted with images through webcams.
2. As they consult the webcam, they have to record the material description of the image into a table and then identify the scales of integration of the landscape, the relationships between the actors identified, and the potential conflicts; students are expected to be familiar with such concepts by the end of the secondary school course.
3. At the same time as or after the observation, students are invited to use abstraction to categorize the issues associated with the landscape studied and to state possible reasons for mobilization and/or proposals for action.

8.3.3 Major Results and Critical Perspective

Figure 8.1 is made up of elements (words and concepts) extracted from students' completed tables. It is based on permanent interactions between immediate and personal contact, representations, information (by webcams), and critical thinking. The results of the surveys show a gradual break with the "matrix narrative of territorial competition in globalisation" (Charpentier et al., 2023) as prescribed in the *Lycée* curricula.

Clearly, two vectors of globalization have been experimented with. These are the processes of integration and ranking of territories. The coastal locations of the webcams chosen have something to do with this: The container ships off the port of Rotterdam and the cruise ship moored in the Norwegian fjords set the students thinking about the flows and circulation of globalization. As for the ranking process of territories, in a more subtle way, it is embodied in the power relationships between centers and peripheries regarding geopolitical aspects (military ship observed in Norway), the co-presence between tourists and inhabitants (deckchairs/sweepers), or the markers of the cultural reception of globalization, the latter being underlined regularly (westernization, conflict of identity). However, the most striking element is the prevalence of environmental risks (artificialization, concreting), social risks (cohabitation, conflicts of use, privatization), and global risks (seabed, submersion) in the objectification phase of the observations.

This can be interpreted as students distancing themselves from the teaching of the organization of the world through globalization in favor of economic production

Table 8.2 Sample of students' table

	Before	During			After
Location	Spontaneous spatial representation of places	Description/Time. What does the webcam show?	Amenities/Conflicts of use/Risks (environment, social or economic)	Location advantages and issues (economic, cultural...) / Scales of integration	Possible reasons for mobilisation and/or proposals for action
Lost Bar- Choeng Mon-Koh Samui Thailand https://www.worldlivecamera.com/fr/webcam/Choeng-Mon-Koh-Samui.html	Beach and coconut trees Beach bar	11 am. Nobody on the beach. Some tables and chairs. Several fishermen's boats	Few infrastructures except a dyke Potential conflicts between tourists and fishermen	The place could attract more tourists	Biodiversity loss: too much tourists could damage environment. Introduction of measures to limit tourism
Sangatte Pas-de-Calais France https://www.visionenvironnement.com/livecams/webcam.php?webcam=sangatte	Migrants hiding out/crossing the Channel Jungle immigration camp	3 p.m. Container ships Large beaches Empty walking/cycling path	Measures to protect Sand Dunes from Erosion. Risk of ship collision and pollution	Safety for migrants and boats	Public image of the location and reputation management. Citizens' expectations of public authorities e.g. local government
Hotel Roth am Strand Westerland Sytl — Schleswig-Holstein Germany https://www.youtube.com/watch?v=HXnlOrHPAMQ	Private beach for hotel guests only (sofas, deck chairs, bars)	11 am, stormy. Low visibility Beach with wicker cabins on sandy beach	German design style. Well known typical north-beach. Privatisation against free access to public beach/coastline Risk of segregation/social tensions (punks, inhabitants)	Worldwide place marketing for happy few	Requirement for free access to the beach

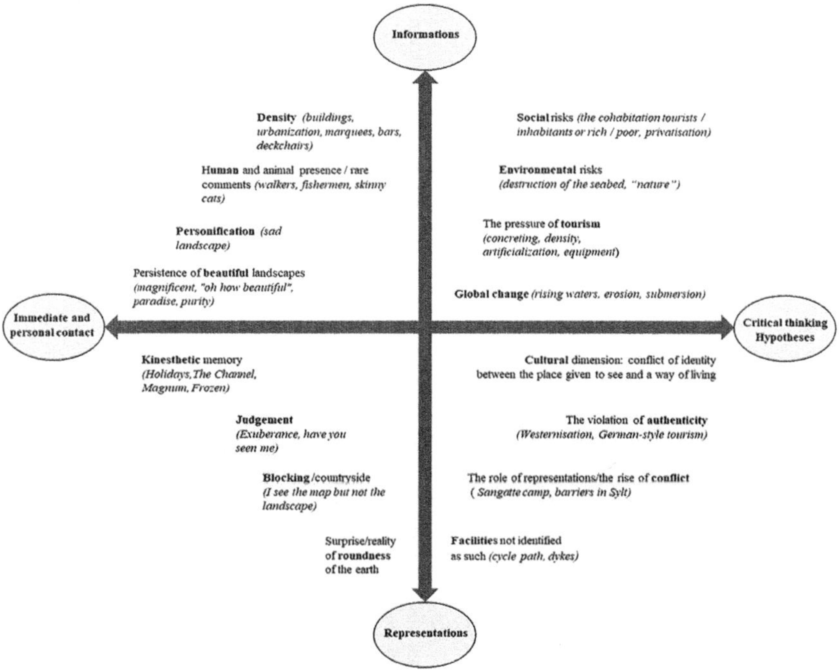

Fig. 8.1 Analysis of the surveys

(*Terminale* curriculum) and, conversely, viewed as their resorting to an alternative reading prism that puts societies back at the heart of the reflection while manipulating the knowledge taught (flows, etc.). It cannot be ruled out that this marked shift in thinking toward anthropization is linked to the use of landscape analysis via webcam; approaching globalization through the eyes of an observer via a webcam refocuses the resulting knowledge on the effects of globalization on societies (Fig. 8.1).

In terms of the process of constructing this knowledge, all the stages form a system. The immediate personal contact with the images relayed by the webcams is strongly structured by the imaginary representation of places and representations. For example, some students were surprised, almost disappointed, when they saw the images of Sangatte (France), not to be able to detect any traces of migrants heading for the United Kingdom in the landscape (flag, camp), or in Sylt (Germany), not to see any punks installed next to the wicker beach benches that dot the seafront to offer shelter from the sand and wind of the North Sea. It did not fit in with their idea of the place, and so it did not fit in with their expectations of the landscape. The interplay between these first two stages can be observed even through the information gathering and reasoned processing phase, as well as the explanatory phase. Again at Sangatte, the coastal tourist facilities (cycle paths, rockfill, benches), although visible in the foreground, as well as the intense maritime traffic in the Channel in the background, went unnoticed and obscured the link between these places and the world through tourist or commercial flows. In this case, in an almost caricatural way, the representations stood in the way of globalization.

Finally, the flexibility of the protocol used (distance learning, free management of investigation time, and connection times) enabled the students to test in practice the accuracy of the knowledge acquired in class. This is the case, for example, with the Earth's roundness. It is very different to learn in class that the Earth is round and to find, when connecting at 5 p.m. to the various survey webcams, that it is impossible at the same time to describe the landscape of Saint Lunaire (France), Koh Samui (Thailand), and Hawaii (USA) because, in two of these places, quite simply, it is night.

8.4 Developing Transferrable Skills Through a Novel and Meaningful Approach to Landscapes

8.4.1 The Culture of Civic Commitment: Promoting Students' Awareness of the Interconnection Between Landscapes and Globalization

When the proposals are pooled, it becomes clear that the natural dimension of the landscape is still highly valued by the students. Interestingly, it proved to be the driving force behind their civic commitment. In the hit parade of landscapes, the Norwegian fjord topped the list as the quintessential landscape, that is, a landscape integrated into globalization but not shaped by it. The ideal of globalization thus becomes a kind of controlled process balancing "natural" elements with the orientation of the territory toward competitive economic production on a global scale. By contrast, Dubai's coastal landscape is the stuff of grievances, from the "disproportionate amount of land reclaimed from the sea" to the privatization of areas devoted to global tourism. In both cases, the myth of the "balanced landscape" produces emancipatory knowledge (Charpentier et al., 2023) in the sense that it encourages students to formulate solutions to optimize the effects of globalization once a problem has been identified. This is the first step toward commitment.

When formulating proposals for action, the ideas of the students followed two main structural trends: either they took a stance against the markers of globalization in the landscape, sometimes in a radical way (financial sanctions, prohibition, destruction), or they took a stance in favor of preserving the remaining sites (limitation, consultation, sanctuary). Finally, more rarely but also more creatively, some proposed the development of new sites to anticipate global change and pre-empt the risks identified in the previous phases of the survey. This type of creative proposal means that the student has integrated the knowledge acquired in class and, above all, gone beyond it through experimentation. Fatalism and the logic of giving up are positions taken by some students (3/15) who do not question certain previous choices of layout (Rotterdam) and advocate selectivity in future layouts.

Another possible interpretation is that there is a heightened awareness of the possibility of collective action, the mechanism of which is fairly simple (Westheimer & Kahne, 2004). The impetus is provided by the spontaneous individual reaction to the discovery of webcam images ("I find this landscape magical"), which the student assumes is shared by others (so do others). From there, it becomes likely that the valued elements of the landscape will be altered by a massive influx of actors with conflicting interests, and, in response, the need for a strong collective stance will arise. The driving force behind this commitment is the emergence of the notion of the common good in the debate on the landscape, in the context of globalization (Pouthier, 2021). The landscape thus becomes a common good in the political sense (Gerber & Hess, 2017), transformed by the spatial logics of globalized players. Moreover, it should be noted that only citizens' initiatives and the State are seen as guarantors of this common good, through their action of negotiation and shared management of the inhabitation of a landscape. The role of the intermediate levels of associations and non-governmental organizations is not regarded as effective.

8.4.2 Improving Critical Thinking in the Digital Age

Finally, it appears that the proposals for collective action that result from the understanding of globalization through landscape analysis involving webcams do pass the test of critical examination when they remain in a virtual space of exchange (Lupatini, 2019). The individual commitment of students is always hypothetical. The way in which a classroom operates cannot be totally assimilated to that of society, and therefore of public debate, and there is a long way to go between taking a stand, defending one's position, and taking action (Leleux et al., 2017). Nonetheless, I would argue that it is necessary to compare the mobilization proposals put forward by each individual to test their ability to argue and their propensity to defend a position that is sometimes ambitious. It, therefore, requires debate, and to do this, video-conferencing, mind-mapping, or virtual and cooperative whiteboard solutions are not enough for high school students. Moreover, the slowness with which distance learning tends to operate, for technical reasons, is at odds with the enthusiasm with which students defend their proposals. In this respect, distance learning only is counterproductive. To be fruitful, the critical exercise requires a face-to-face debate, making the most of all modes of expression, including non-verbal communication (gestures, body and facial language, etc.).

Particular attention should also be paid to the paucity of reflection on the use of webcams as a tool for analysis. At the time of the surveys or in the debate phase that followed, few students asked questions about the legal framework or simply about the divergent interests of the players (sponsors, service companies, the public) in the global network for the dissemination of images captured in public spaces and thereby publicized. As a result, there is a vast untapped field of investigation for teachers in the context of media education (E.M.I.) in relation to the institutional commission and the transformations of societies linked to the omnipresence of images and digital technology in contemporary times.

8.5 Conclusion

At the end of this chapter, it appears that the study of landscapes, renewed by webcams, significantly facilitates the understanding of the societal challenges posed by globalization. More than just a pedagogical tool, this approach sets students on the path to building cross-disciplinary skills that will undoubtedly benefit the civic education of students-dwellers. It deconstructs the relationship between learning and digital skills, which is still heavily influenced by inequalities (access, use), and rebuilds it in such a way that each student can develop his or her autonomy in going beyond knowledge and, by taking part in public debate, make intelligible proposals for collective action to meet societal challenges. Students' civic commitment in the classroom mirrors that of society as a whole via online multi-actor commitment platforms, including the call for citizens' initiatives Mon projet pour la planète (126,546 participants, 420 eligible projects, €3 million between May 2017 and May 2018) or le Mouvement One Planet initiated in 2017 (6 Summits, 40 initiatives, 3000 researchers, 153 governments).

References

Berque, A. (1990). *Médiance, de milieux en paysages*. GIP Reclus.

Biaggi, C. (2015). Habiter, concept novateur dans la géographie scolaire ? *Annales de Géographie, 704*(4), 452–465. https://doi.org/10.3917/ag.704.0452

Boullier, D. (2019). VI. Rendre le numérique habitable : l'habitèle. In Y. Calbérac, O. Lazzarotti, J. Lévy, & M. Lussault (Eds.), *Carte d'identités L'espace au singulier* (pp. 151–174). Hermann. https://doi.org/10.3917/herm.povoa.2019.01.0151

Brunet, R., Ferras, R., & Théry, H. (2005). *Les Mots de la géographie : dictionnaire critique* (3e éd.). La Documentation Française.

Charpentier, P., Leininger-Frézal, C., Genevois, S., Leroux, X., Thémines, J., & Vergnolle-Mainar, C. (2023) . Chapitre 1. Contextes et conditions de pratiques de savoir émancipatrices à l'École, à partir du cas de la géographie scolaire française. In Buttier, J. et Panagiotounakos, A. (dir.), *Des savoirs pour agir sur le monde Quels apprentissages des élèves face aux enjeux contemporains ?* (pp. 27–51). Presses universitaires de Grenoble. https://doi.org/10.3917/pug.butti.2023.01.0027

Genevois, S. (2020). *Le numérique dans l'enseignement-apprentissage de la géographie : quels apports, quels enjeux ?* Cenesco-Cnam. https://www.cnesco.fr/wp-content/uploads/2021/02/210218_Cnesco_Genevois_Numerique_Geographie.pdf

Gerber, J.-D., & Hess, G. (2017). From landscape resources to landscape commons: Focussing on the non-utility values of landscape. *International Journal of the Commons, 11*(2), 708–732. https://doi.org/10.18352/ijc.745

Grataloup, C. (2023). Introduction. Le temps du Monde. *Géohistoire de la mondialisation - 3e éd. Le temps long du monde.* (pp. 9–18). Armand Colin. https://shs.cairn.info/geohistoire-de-la-mondialisation%2D%2D9782200636388-page-9?lang=fr.

Heimberg, C. (2011). L'éducation à la citoyenneté à Genève et en Suisse romande : comment faire valoir sans prescrire ? *Vierteljahrsschrift Für Wissenchaftiche Pädagogik, 87*, 520–532.

Joublot Ferré, S. (2021). L'enquête paysagère, une méthode d'intelligibilité politique du monde ? *Projets de Paysage, 24*. https://doi.org/10.4000/paysage.20149

Lazzarotti, O. (2014). *Habiter le monde : La documentation photographique. Dossier n° 8100*. La documentation Française.
Leleux, C., Rocourt, C., & Lantier, J. (2017). *Éducation à la philosophie et à la citoyenneté : didactique et séquences*. De Boeck Education.
Lévy, J., & Lussault, M. (2003). *Dictionnaire Le géographie et de l'espace des sociétés* (1re éd.). Belin.
Lévy, J. (2007). La mondialisation : un événement géographique. *L'Information géographique, 71*(2), 6–31. https://doi.org/10.3917/lig.712.0006
Lewis, M. W., & Duncan, J. S. (1992). The city as text: The politics of landscape interpretation in the Kandyan Kingdom. *Geographical Review, 82*(3), 349. https://doi.org/10.2307/215367
Lupatini, M. (2019). Controversies on public space management projects in Geography teaching to educate pupils as dynamic citizens. *Area, 51*, 653–651. https://doi.org/10.1111/area.12523
Lussault, M. (2019). Entrée en matière La spatialité d'abord. In Y. Calbérac, O. Lazzarotti, J. Lévy, & M. Lussault (Eds.), *Carte d'identités L'espace au singulier* (pp. 7–24). Hermann. https://doi.org/10.3917/herm.povoa.2019.01.0007
Pols, J. (2018). Fabuleuses webcams : Regards actifs et technologies invisibles. *Réseaux, 207*(1), 65–94. https://doi.org/10.3917/res.207.0065
Pouthier, F. (2021). Les parcs naturels régionaux, générateurs de « communs » dans les territoires. *Nectart, 12*(1), 114–125. https://doi.org/10.3917/nect.012.0114
Programme de l'enseignement d'histoire-géographie de la classe de seconde générale et technologique, de la classe de première de la voie générale et de la classe de première de la voie technologique. (2019). Ministère de L'Education Nationale, de L'Enseignement Supérieur et de la Recherche. https://www.education.gouv.fr/bo/19/Special1/MENE1901577A.htm
Programmes d'enseignement de l'école élémentaire et du collège. (2015). Ministère de L'Education Nationale, de L'Enseignement Supérieur et de la Recherche. https://www.education.gouv.fr/au-bo-special-du-26-novembre-2015-programmes-d-enseignement-de-l-ecole-elementaire-et-du-college-3737
Sawyer, C. F., Butler, D. R., & Curtis, M. (2010). Using webcams to show change and movement in the physical environment. *Journal of Geography, 109*(6), 251–263. https://doi.org/10.1080/00221341.2010.506924
Sgard, A. (2015). Chapitre 9. La controverse publique comme objet et dispositif d'apprentissage en géographie : questionner la frontière nature-culture. *Sciences de la nature et de la société dans une école en mutation Fragmentations, recompositions, nouvelles alliances ?* (pp. 113–124). De Boeck Supérieur. https://shs.cairn.info/sciences-de-la-nature-et-de-la-societe-dans-une-ec%2D%2D9782807301764-page-113?lang=fr.
Westheimer, J., & Kahne, J. (2004). What kind of citizen? The politics of educating for democracy. *American Educational Research Journal, 41*(2), 237–269. https://doi.org/10.3102/00028312041002237

Elisabeth Dury is a geographer affiliated with the HABITER Research Team (EA 2076) at the University of Reims. Her main research field is Berlin, where she currently focuses on issues related to urbanity and borders, territorial marketing, as well as citizen mobilization and participation. She is also an agrégée professor of history and geography taught in German within the ABIBAC program at the Jean Jaurès High School in Reims. In addition, she serves as a lecturer at URCA and a member of competitive exam juries, which reflects her interest in the didactics of geography.

Chapter 9
Teaching Geo-space Through Socioscientific Issues: When Earth Geography Meets Space Geography

Georgios Bampasidis and Apostolia Galani

9.1 Connecting Geography with Planetary Science Socioscientific Issue Contexts

Geography is undoubtedly one of the most important disciplines in the school curriculum as it has the advantage of clarifying our planet's complexity. It facilitates our understanding of Earth's dynamics, bringing order in a chaotic world (Bonnett, 2023). It also investigates the connections between the natural environment and society, which is vital in studying and decoding the mysteries of our world.

Geography also examines the humans impact on the planet by studying how activities, such as urbanization, deforestation, and industrialization, may modify ecological conditions (Arthur, 2021), and it examines how such changes can affect sociocultural systems. In this context, Geography analyzes human-induced spatial changes in terms of sustainability and resilience, encourages ethical consideration of the implications of human–environment interaction, and addresses global issues such as climate change, resource depletion, and justice concerns (Galani, 2016; Karatza et al., 2020).

From the beginning of the space era, Astronomy has progressed significantly beyond the observation of the starry sky. Its new subfield, Planetary Science, has significantly advanced our knowledge about the solar system bodies. Borrowing techniques and methods from the related disciplines, such as Geology or Space Physics, Planetary Science examines the formation and evolution of the planetary

G. Bampasidis (✉) · A. Galani
Department of Pedagogy and Primary Education, National and Kapodistrian University of Athens, Athens, Greece
e-mail: gbabasid@primedu.uoa.gr; ligalani@primedu.uoa.gr

M. Hardouin, C. Leininger-Frézal (eds.), *Curricular Changes in Geography Education*, International Perspectives on Geographical Education,
https://doi.org/10.1007/978-3-031-96551-7_9

bodies as well as their distinct characteristics. Lewis (2004) states that Planetary Science stands as a great example of interdisciplinarity, as it unifies the fundamental sciences by dismantling the artificial boundaries that have constrained our understanding of the world. From this perspective, both Planetary Science and Geography share a common reference point in exploring our planet and bodies of the solar system through the framework of the geospheres. Researchers have increasingly integrated Planetary Science into Geography, emphasizing the study of geological processes that govern celestial bodies rather than merely cataloging their features (McSween et al., 2019).

In the field of education, Astronomy is mainly incorporated into curricula as supplementary content to enhance other disciplines rather than as an independent field of study. The International Astronomical Union provides a comprehensive summary of the astronomical content included in the curricula of 89 countries worldwide (https://astro4edu.org/). As reviewed by Bampasidis (2024), several trends emerge: (1) Astronomy is included in both Science and Geography courses in 40 countries; (2) In most countries, Astronomy is not offered as a standalone subject: It is excluded from the curriculum in 77 countries (86.52%) and taught independently in only 12 (13.48%); (3) Astronomy is more frequently included in Science (88.64%, or 78 countries) than in Geography curricula (51.72%, or 45 countries). The European Space Agency (ESA, 2002) highlighted the inflexibility of curricula, where space-related subjects are often elective. To foster interest in space, ESA has founded the European Space Education Resource Offices in various countries. NASA and science museums have developed similar space educational programs. The findings also show that when Astronomy is included in science courses, it reflects its connection with natural phenomena, while its inclusion in Geography emphasizes space as a component of the natural world, often without addressing the broader social implications of Science.

As a response to this need, this chapter describes a proposal for a reconceptualized Geography Curriculum that embeds topics from Planetary Science through the context of socioscientific issues (SSI). This interdisciplinary approach aims to enhance student's awareness about the current global challenges and their connection with human activities. The chapter is organized as follows. Section 9.1 introduces the connection between school Geography and Planetary Science. Section 9.2 outlines the procedural aspects of the approach, describing the methodology and presenting the selected contexts in Geography, Planetary Science, and SSI. Section 9.3 describes how the curriculum is enhanced through interdisciplinary themes and SSI context. The final section (Sect. 9.4) presents the pilot implementation of this proposal and discusses the conclusions.

9.2 Terrestrial Geospheres and Planetary Frontiers: A Guide to Reconceptualizing the Geography Curriculum

In this section, we describe the methodology used to develop the reconceptualized Geography Curriculum. The approach began by selecting key concepts from Planetary Science to integrate into the existing Geography curriculum. The authors have created new educational material for fifth- and sixth-grade elementary students, as well as junior high school Geography Curricula (Bampasidis, 2024).

9.2.1 *Methodological Approach*

The reconceptualization of the Geography Curriculum is based on the geosphere model, commonly used when studying the environment within Physical Geography (Arbogast, 2020). According to this model, the Earth is divided into four spheres: the Atmosphere, the Hydrosphere, the Lithosphere, and the Biosphere. The four geospheres are interconnected and influence each other. The geosphere model provides an educational framework to understand the interactions among Earth's systems. In our approach, this model enables students to make meaningful comparisons between terrestrial and extraterrestrial geophysical processes. The reconceptualized Curriculum adopts the framework of SSI, integrates Planetary Science topics, and addresses key environmental and societal challenges shaped by human activity. Our approach addresses major global concern topics such as the greenhouse effect, climate change, water distribution, hydrogen production, rare earth elements (REEs), the potential for life beyond Earth, and space colonization, informed by insights from space research.

9.2.2 *Selection of Topics in Geography and Astronomy and Their Common Themes*

We first identified key themes within both Geography and Planetary Science that align with the objectives of SSI education. We have had to review the existing Geography curriculum to identify the specific thematic areas where the inclusion of Planetary Science concepts could support the understanding of environmental and societal dynamics. In order to foster a deeper understanding of the dynamic relationship between the environment and society, our reconceptualized Geography Curriculum integrates insights from space research with pressing global issues, such as the greenhouse effect, climate change, and the spatial distribution of water resources. It also includes debatable topics, such as hydrogen production, mineral economics of REEs, the quest for life beyond Earth, and the possibilities of space colonization. These topics were selected for their ability to bridge Earth-based studies and Planetary Science, fostering a holistic understanding of these issues in various contexts and promoting the interaction between physical processes and societal impacts.

9.3 Linking the Reconceptualized Curriculum with Environmental and Societal Changes

Integrating SSI contexts from Planetary Science into school Geography involves embedding the study of planetary systems, including the Earth, within the broader social, environmental, and ethical contexts that Geography typically explores. The SSI approach helps students perceive Planetary Science contexts not merely as a collection of physical and astronomical phenomena but as contexts with significant implications for human society, environmental sustainability, and global challenges. SSI education helps students (1) explore the ethical dimensions and values embedded in scientific decision-making, directly connecting these concepts to their lives and their understanding of the world; (2) create an environment that promotes active community involvement, both within and beyond the classroom, which is crucial for cultivating responsible citizenship (Zeidler, 2014); (3) develop decision-making skills, enhance scientific literacy, support intellectual and moral growth, and foster community engagement-outcomes that are relevant on both local scale and global scales (Chowdhury et al., 2020); (4) engage with significant societal and global issues, deepening their understanding of key concepts and empowering them to make well-informed decisions (Sadler, 2004); and (5) critically consider the ethical, environmental, and social dimensions of scientific knowledge and its application (Kumar et al., 2024).

To ensure that the abovementioned selected topics effectively align with SSI, we evaluated each topic to determine whether it adequately addresses social, economic, or political dimensions. Most topics encompass these aspects, providing opportunities to discuss the broader implications of scientific knowledge and its application in societal contexts. We also enriched the content of the topics that lack the previous criterion by incorporating additional elements that emphasize social and political relevance. For example, we deepened the exploration of certain topics to highlight socio-political factors, such as the economic impacts of climate change on various regions.

The main SSI concepts selected by following the described methodology are as follows:

(a) *Greenhouse Effect and Climate Change*

Research findings from space missions that explored rocky planets and gas-giant moons helped us understand the greenhouse effect and climate change. Indeed, remote observations of the thick Venusian atmosphere unveil a runaway greenhouse effect and give warnings for Earth's potential climatic future. In fact, studies of

Venus's atmosphere underscore the urgency of mitigation efforts, highlighting the severe impacts of unchecked greenhouse gas emissions (Svedhem et al., 2007). Venus can serve as a comparative example in an educational context too, since students can better understand the consequences of greenhouse gas emissions and the need to take action on Earth.

(b) *Water Spatial Distribution*

Another interesting result of space exploration is the discovery of water ice on the Moon (Li et al., 2018) and Mars (Gulick, 2001). These detections have led to discussions about water resources in space and their implications for future space missions as well as human colonization. The proposed curriculum incorporates this knowledge to draw attention to the importance of water distribution as a critical resource on Earth and water sustainable management.

(c) *Hydrogen Production and Renewable Energy*

Space exploration is strongly related to alternative energy sources, especially hydrogen production and storage. Astronauts have conducted many experiments on human spaceflight missions to examine the potential of *in-situ* extracting hydrogen from water ice. These efforts demonstrate that similar hydrogen production techniques can be applied on Earth, highlighting the connection between space technology and renewable energy needs (Sanders, 2015). This knowledge is directly relevant to Earth's need for clean energy, highlighting hydrogen's potential as a future energy carrier to reduce greenhouse gas emissions. Our curriculum adapts such discussion to unveil the way that space technologies can provide sustainable energy solutions on Earth and strengthen the idea of Earth as an interconnected system influenced by natural and human factors.

(d) *Rare Earth Elements and Mineral Economics*

Mining REEs on the Moon and asteroids became a topic of discussion since the *in-situ* study of the chemical composition of their surfaces. Such mining, if possible, could revolutionize the global supply of these expensive resources. Yet, such discussions have triggered significant ethical and geopolitical concerns, as space mining could drastically affect global economics and resource distribution (Gregg, 2021). With the growing demand for REEs in high-tech and green technologies, space mining offers new perspectives on the economics of mineral resources. The proposed curriculum incorporates these insights, debating the socio-economic and environmental implications of REE extraction both on Earth and in space and evaluating the geopolitical and ethical challenges of space mining and its potential to alleviate or worsen resource scarcity on Earth.

(e) *Potential for Life Beyond Earth*

The quest for life in space explores a twofold problem. It tries to understand the factors that permit life to emerge and sustain in extraterrestrial environments and work on space colonization plans, providing insights into sustainability and human impact on other worlds (Torres et al., 2021). Missions to Mars, the Moon, and giant planets' moons such as Europa and Enceladus focus on detecting signs of life and addressing the challenges of human living in harsh conditions, raising ethical questions about altering extraterrestrial environments.

Integration of these SSI concepts in the Geography Curriculum offers a real holistic view of how human activities, changes to the environment, and societal transformation interact. This deepens knowledge in Earth and Space Sciences through encouraging critical thinking in matters that could present challenges in the future.

9.3.1 Expanding the Geospheres: Enhancing Geography Lessons Through a Focus on Interconnected Systems

Below we outline how the four geospheres may be taught through both geography and planetary science. We aim to deepen students' understanding of the geospheres and their broader planetary context by comparing Earth's geospheres and those of other celestial bodies. This approach enhances key geographical concepts and, by integrating insights from space exploration, makes the study of geospheres a dynamic and interdisciplinary experience. Additionally, through the study of geospheres, we address key SSI, such as climate change, encouraging students to critically engage with the challenges and ethical considerations arising from the intersection of human activities and planetary environments.

Thematic Approach #1: Flying Into the Atmosphere by Using Solar System Analogs

Geography textbooks typically focus on the terrestrial atmosphere's temperature profile, portraying its stratified structure. On the other hand, the planetary atmospheres are the most extensively studied environments throughout the solar system, by either *in-situ* space missions or remote observations. Our methodology emphasizes analogies between the Earth's atmosphere and the other planetary atmospheres in the solar system.

Figure 9.1 illustrates how the reconceptualized Geography curriculum integrates interdisciplinary themes related to the atmosphere. The upper side of the diagram represents the terrestrial atmosphere focusing on traditional Geography such as meteorology, the ozone layer, and the interaction of the atmosphere with solar radiation. Other topics illustrated are the vertical temperature profile, the study of atmospheric pollution, and its environmental and health effects. The lower side focuses on planetary atmospheres, highlighting topics of common interest such as photochemistry, which examines chemical reactions triggered by light, the origin and evolution of planetary atmospheres, which explores how these atmospheres form and change over time, and atmospheric replenishment, which looks at the addition and removal in planetary atmospheres.

As Figure 9.1 depicts, the intersection of these areas captures the reconceptualized curriculum learning concepts as we design it. It emphasizes the structure of

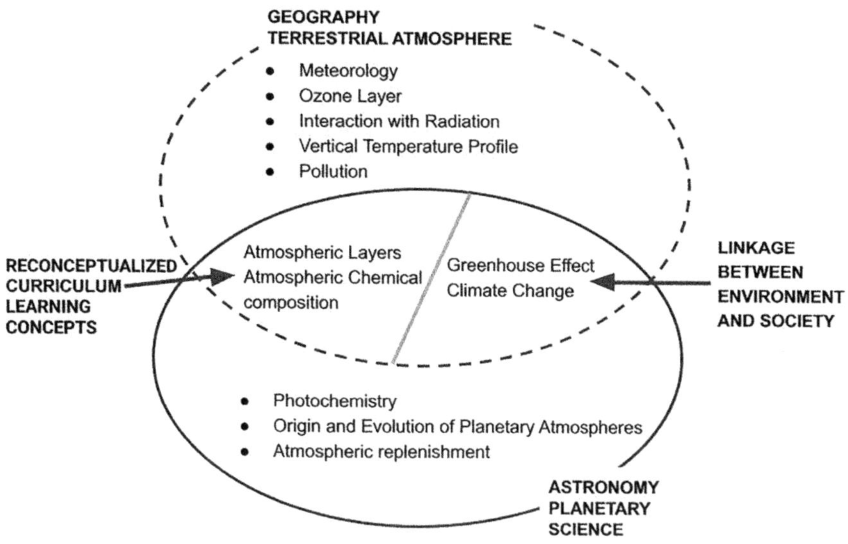

Fig. 9.1 Integration of Planetary Science with Geography in the study of the atmosphere. Concepts from these two disciplines are combined to develop a curriculum that emphasizes the connection between environmental phenomena and societal effects and extends the depth of knowledge about Earth's atmosphere to those of other planets

atmospheric layers, comparing Earth's atmosphere to those of other planets and comparing the chemical composition of different planetary atmospheres. It also addresses the greenhouse effect, examining how greenhouse gasses affect planetary temperatures and drawing comparisons of this phenomenon among the rocky planets and gas giant moons. Another key concept is climate change, exploring how atmospheric processes drive climate change on Earth and how these insights can apply to other planets.

Our approach also discusses how atmospheric changes on Earth and other planets have significant implications for human societies. When students understand the greenhouse effect and climate change through Planetary Science, they will become more aware of the urgent need for action on Earth. In addition to the previously SSIs, the curriculum is designed to address the necessity for sustainable environmental practices and the importance of understanding atmospheric processes across different planetary bodies, encouraging students to respond to Earth's challenges.

Thematic Approach #2: Diving into the Hydrosphere: Making Connections with the Solar System Water Sources

In the context of school Geography, the conventional curriculum typically associates the hydrosphere with the study of Earth's water dynamics and the description of aqueous reservoirs. However, planetary research has revealed that Earth is not the only celestial body with water or liquid deposits. Recent findings suggest that water or other liquids may exist, or have existed, on the surfaces of various celestial bodies in our solar system. Additionally, liquids are found within the gaseous clouds of the giant planets and may exist as oceanic deposits beneath the icy crusts of their moons.

Figure 9.2 illustrates how the reconceptualized Geography curriculum integrates interdisciplinary themes by focusing on the hydrosphere. In the top side of the Figure, the "traditional" Geography concepts of the terrestrial hydrosphere are outlined, such as the water cycle, the oceanic water circulation, and the water pollution and the impact on ecosystems. Figure 9.2 also presents, as traditional concepts, the study of aquatic ecosystems and water conservation, emphasizing the management of resources in a sustainable manner to ensure availability for upcoming generations. The other side of Figure 9.2 shows the selected topics from Astronomy and Planetary Science that focus on extraterrestrial water deposits. This includes the investigation of subsurface salty oceans on moons such as Europa and Enceladus, the tidal heating procedure that could maintain liquid water beneath icy surfaces, and the potential extraterrestrial origin of liquids in different environments across the solar system.

As shown in Figure 9.2, the basic concepts of the reconceptualized Geography Curriculum about the hydrosphere focus on the water triple point and its phase diagram. It illustrates the states of water and examines the water distribution across planetary bodies. It also expands water's importance for Earth, space exploration, and colonization by considering it as an energy resource for hydrogen production. Thus, through water's critical role in sustainability and socio-economic development, the integration of these concepts in the curriculum forges the connection

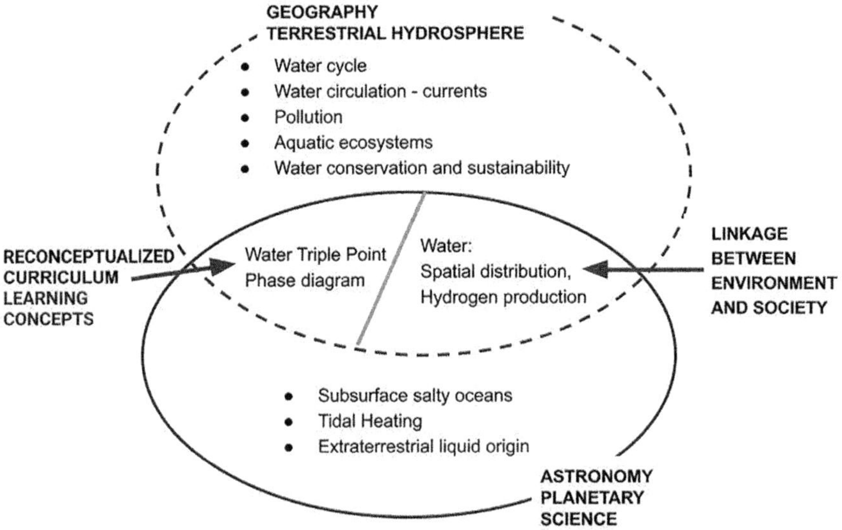

Fig. 9.2 Integration of Planetary Science with Geography in the study of the hydrosphere. Concepts from these two disciplines are combined to develop a curriculum that emphasizes the connection between environmental phenomena and societal effects and extends the depth of knowledge about Earth's hydrosphere to those of other rocky planets and moons

between environment and society. This approach helps students understand the crucial and advanced role of water in a broader planetary context.

By connecting these themes, the curriculum fosters a comprehensive understanding of the hydrosphere, both on Earth and in space, enhancing students' appreciation of water's role in sustaining life and shaping environments. This interdisciplinary perspective broadens their knowledge of Geography and Planetary Science and encourages consideration of the socio-environmental implications of water use and conservation in a global and cosmic context.

Thematic Approach #3: Digging the Lithosphere: Learning About Terrestrial-Like Planetary Landscapes

Traditional school Geography presents Earth's surface structures by focusing on their description and the processes that shape them, viewing the Earth's surface as a dynamic system undergoing constant evolution.

Our approach extends the concept of the lithosphere beyond our planet by incorporating Planetary Science. This allows for a comparative study of the inner planets and natural satellites, examining common geomorphological characteristics with their terrestrial counterparts.

Figure 9.3 illustrates how the reconceptualized Geography curriculum integrates interdisciplinary themes with a focus on the lithosphere. One side of the Figure represents the terrestrial lithosphere, covering traditional Geography and Earth Sciences topics, such as plate tectonics, continental drift, Mineralogy, and geological history. The other side focuses on Astronomy and Planetary Science, highlighting geological features and processes beyond Earth. This includes the study of impact craters formed by celestial collisions, cryovolcanism on icy moons where volatile substances such as water, ammonia, or methane erupt instead of molten rock; regolith and tholins found on planetary surfaces, with regolith referring to the loose, fragmented material covering solid rock, and tholins being complex organic molecules that form in planetary atmospheres; and polar ice caps, which provide insights into environmental conditions across different planets.

The intersection of these areas represents the reconceptualized Curriculum learning concepts, integrating knowledge from terrestrial lithosphere studies and Planetary Science. The curriculum examines geological characteristics on Earth and compares them with those found on other planets and moons exploring surface features, planetary interior, and volcanic activity, including traditional volcanism and cryovolcanism.

Another concept is this of rare-earth elements and mineral economics, discussing the economic importance of minerals on Earth and in space, and the potential for mining resources on other planets and moons. This discussion also touches on the ethical and geopolitical implications of extraterrestrial mining, linking to resource management challenges on our planet. By integrating these topics, the Curriculum emphasizes the linkage between environment and society by exploring how geological processes and mineral resources, both on Earth and in space, are not just scientific phenomena but also have profound impacts on human societies. For instance, understanding the distribution and economics of rare-earth elements is

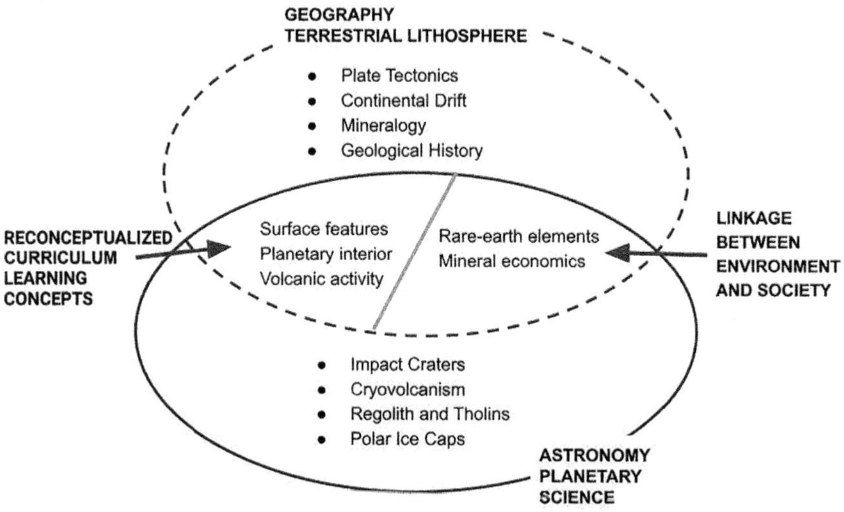

Fig. 9.3 Integration of Planetary Science with Geography in the study of the lithosphere. Concepts from these two disciplines are combined to develop a curriculum that emphasizes the connection between environmental phenomena and societal effects and extends the depth of knowledge about Earth's lithosphere to those of other rocky planets and moons

critical for industries and technologies on Earth and could shape future space exploration and exploitation.

By connecting these themes, the curriculum provides a comprehensive understanding of the lithosphere, both on Earth and in space, encouraging students to consider the socio-environmental and economic implications of geological phenomena in a global and cosmic context.

Thematic Approach #4: Is There Anybody Around?: Engaging Geography with the Research for Extraterrestrial Life Traces

Examining the geospheres within the solar system yields a wealth of information relevant to the quest for conditions conducive to extraterrestrial life and explores the origins of life on Earth. This approach extends the Biosphere's margins, studying the spatial distribution of biogenic elements on the Earth and beyond.

Figure 9.4 illustrates how the reconceptualized Geography curriculum integrates interdisciplinary themes of the Biosphere. One side of the Figure represents the terrestrial Biosphere, highlighting traditional topics, such as the study of ecosystems and their classification based on environmental conditions and species diversity. Figure 9.4 emphasizes the connection between ecosystem and organism diversity, showing how diverse ecosystems contribute to biodiversity. It further looks into ecological balance conservation, highly crucial in the sustainability of ecosystems that support life, and identifies regions of high ecological interest important for biodiversity conservation and sustainable uses of resources. The lower side of Figure 9.4 focuses on life beyond Earth, a concept derived from Planetary Science.

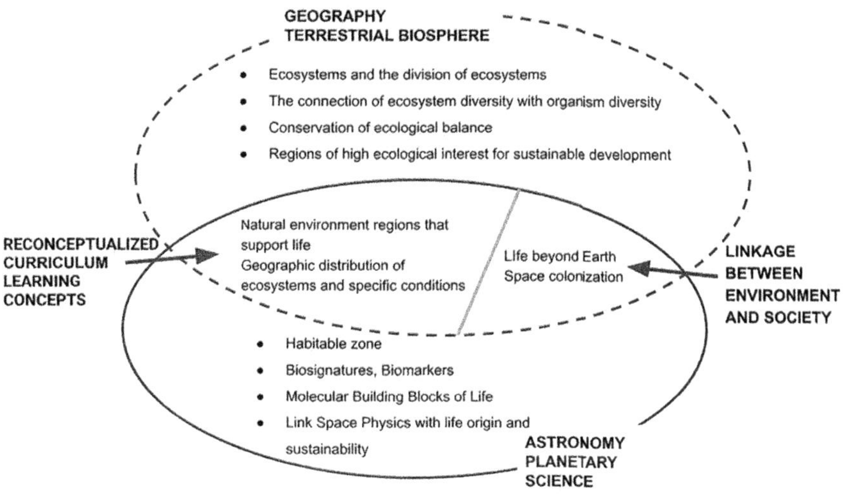

Fig. 9.4 Integration of Planetary Science with Geography in the study of the biosphere. Concepts from these two disciplines are combined to develop a curriculum that emphasizes the connection between environmental phenomena and societal effects and extends the depth of knowledge about Earth's biosphere too

This habitable zone refers to the region around a star with possible proper conditions for life. It provides an adequate temperature range to allow for liquid water. These concepts also include the research for biosignatures and biomarkers as possible life detectors on other planets and the exploration of the molecular building blocks of life. Finally, it discusses how Space Physics might impact life beyond Earth, studying factors such as radiation and microgravity.

The reconceptualized Curriculum learning concepts lay in the intersection of these areas where terrestrial biosphere studies and planetary science are integrated. It focuses on regions that support life, both on Earth and potentially on other planets, and explores the geographic distribution of ecosystems, comparing the Earth's diverse ecosystems with potential ecosystems that could exist on other celestial bodies.

The curriculum, by addressing the concepts of life beyond Earth and space colonization, encourages students to consider the possibilities and challenges of finding extraterrestrial life or/and establishing human colonies outside Earth. Thus, the ethical and practical implications of colonizing other planets arise. When understanding the conditions necessary for life, students consider the challenges of sustaining ecosystems both on Earth and in potential space colonies and enhance their understanding of the biosphere, both on Earth and in a broader cosmic context. This interdisciplinary approach enriches students' knowledge of Geography and Planetary Science, encouraging them to consider the socio-environmental and ethical implications of sustaining life in different terrestrial and extraterrestrial environments.

9.4 Conclusion

Our approach was piloted twice during the MSc course "Remote Sensing and Astronomy: Exploring Questions about the Environment and the Universe" at the National & Kapodistrian University of Athens during the spring semesters of 2020–21 and 2022–23. Students were satisfied with the educational material and commented that it could be readily adapted for classroom use. They appreciated the coherence of the selected concepts, which are presented here, though many noted that further personal study was needed for deeper mastery and praised the comparative approach between Earth and other planets (Bampasidis et al., 2023; Bampasidis, 2024). This feedback supports the curriculum's ability to bridge Earth Science and Planetary Science, fostering critical thinking and preparing students for global and cosmic challenges.

This chapter outlined the integration of Planetary Science into the Geography curriculum through a framework that compares Earth's geospheres with those of other planets. By emphasizing these interlinked subsystems, the curriculum reinforces key geographical concepts while extending them to a planetary perspective. This approach enables students to contextualize the wider significance of environmental processes and their socio-economic consequences. The interdisciplinary approach presented here incorporates SSI to highlight real-world topics, such as climate change, water distribution, resource management, and the potential for life existence beyond Earth.

By seeking analogies between terrestrial and extraterrestrial environments, this curriculum fosters a comprehensive understanding of planetary systems functions and interactions, encouraging students to think critically about the future of our planet. Also, ethical issues would be discussed regarding the contamination of other worlds via space exploration and the sustainability of long-term human activities, providing a holistic view of the challenges humanity faces on Earth and beyond.

The approach we present in this chapter highlights the unifying power of education. It suggests a way for teachers to integrate modern Astronomy into Geography courses. By integrating Astronomy's cosmic perspective with Earth-centered Geography, students are inspired to become responsible global citizens, equipped with a deeper appreciation of our planet's value, and preparing themselves to successfully address the social and societal challenges of the future. To conclude, this novel Geography curriculum offers an educational experience that prepares students to engage thoughtfully with the complexities facing our planet and the universe, emphasizing the importance of sustainability, ethical considerations, and global cooperation.

References

Arbogast, A. F. (2020). *Discovering physical geography* (4th ed.). Wiley.

Arthur, J. (2021). Impact of geography on adaptation for the future sustainability of human society on earth. *Open Journal of Social Sciences., 9*, 188–217. https://doi.org/10.4236/jss.2021.93013

Bampasidis, G. (2024). (In greek) *Didactic transformation of scientific concepts of astronomy and remote sensing to promote learning: From planetary environments to everyday educational practice* (Postdoctoral research). National & Kapodistrian University of Athens. https://pergamos.lib.uoa.gr/uoa/dl/object/3412901.

Bampasidis, G., Galani, A., Lambrinos, N., & Parcharidis, I. (2023). (In greek) Astronomy in school education: An absence that matters. *13th Conference of ENEFET*. Ioannina, 10–12 November 2023.

Bonnett, A. (2023). *What is geography?* (2nd ed.). Rowman & Littlefield.

Chowdhury, T. B. M., Holbrook, J., & Rannikmäe, M. (2020). Socioscientific issues within science education and their role in promoting the desired citizenry. *Science Education International, 31*(2), 203–208.

ESA. (2002). *A response to TEACH SPACE 2001*. Retrieved August 31, 2024, from https://www.esa.int/Science_Exploration/Human_and_Robotic_Exploration/Education/A_response_to_TEACH_SPACE_2001

Galani, L. (2016). Geo-literacy as the basis of the building of cultural identity. *European Journal of Geography, 7*(1), 17–23.

Gregg, J. (2021). Space mining. In *The cosmos economy*. Copernicus. https://doi.org/10.1007/978-3-030-62569-6_17

Gulick, V. C. (2001). Origin of the valley networks on Mars: A hydrological perspective. *Geomorphology, 37*(3–4), 241–268.

Karatza, A., Galani, L., & Parcharidis, I. (2020). Teaching "Climate Change and Glaciers" using ESA-Eduspace material. *Current Journal of Applied Science and Technology, 39*(15), 68–78. https://doi.org/10.9734/cjast/2020/v39i1530718

Kumar, V., Choudhary, S. K., & Singh, R. (2024). *Environmental socio-scientific issues as contexts in developing scientific literacy in science education: A systematic literature review*. Social Sciences & Humanities Open, Volume 9, 100765, ISSN 2590-2911. https://doi.org/10.1016/j.ssaho.2023.100765.

Lewis, J. S. (2004). *Physics and chemistry of the solar system* (2nd ed.). Academic Press.

Li, S., Lucey, P. G., Milliken, R. E., Hayne, P. O., Fisher, E., Williams, J. P., et al. (2018). Direct evidence of surface exposed water ice in the lunar polar regions. *Proceedings of the National Academy of Sciences, 115*(36), 8907–8912.

McSween, H. Y., Moersch, J. E., Burr, D. M., Dunne, W. M., Emery, J. P., Kah, L. C., & McCanta, M. C. (2019). *Planetary geoscience*. Cambridge University Press.

Sadler, T. D. (2004). Informal reasoning regarding socioscientific issues: A critical review of the research. *Journal of Research in Science Teaching, 41*, 513–536.

Sanders, G. B. (2015). Space resource utilization: Technologies and potential synergism with terrestrial mining. *Canadian Institute of Mining 2015 Convention* (No.JSC-CN-33117).

Svedhem, H., Titov, D., Taylor, F., et al. (2007). Venus as a more Earth-like planet. *Nature, 450*, 629–632. https://doi.org/10.1038/nature06432

Torres, O. A. C., Peters, T., Seckbach, J., & Gordon, R. (Eds.). (2021). *Astrobiology: Science, ethics, and public policy*. Wiley.

Zeidler, D. L. (2014). Socioscientific issues as a curriculum emphasis: Theory, research, and practice. In *Handbook of research on science education* (Vol. II, pp. 697–726). Routledge.

Georgios Bampasidis is an Educational Consultant for Natural Sciences at the Directorate of Public Secondary Education of Athens/Greece and a postdoctoral researcher at the National & Kapodistrian University of Athens. He specializes in science education, particularly the integration of digital technologies and interdisciplinary approaches. His work spans teacher training, STEM education and Astronomy research. His current research interests focus on the use of adaptive learning tools, inquiry-based teaching methodologies, and the application of educational robotics and astronomy in science education.

Apostolia Galani is a Professor in the field of Didactics of Geography at the National and Kapodistrian University of Athens (NKUA). She has extensive experience coordinating numerous European projects, including Erasmus+ and Horizon initiatives, focusing on themes such as the Teaching of Geography, Astronomy, STEM education, and Climate Change. Additionally, she has contributed her expertise to World Bank projects related to education.

Part III
New Subjects to Learn

Chapter 10
Training Teachers to Teach Geopolitics Subjects—A Comparative Study of Israel and Germany

Tal Yaar-Waisel and Sebastian Wolff-Seidel

10.1 Introduction

Geopolitics is the interdisciplinary study that combines geography and politics, examining the impact of geographical factors on political decisions and vice versa (O'Reilly, 2019). Geopolitical science is based on multidisciplinary knowledge and expresses high-order thinking of understanding complex processes and realities, for example, the existence of borders, migration between countries, use of cross-border natural resources, international rivers, national minorities, and more. Geopolitical issues contain the basic questions about our existence and our invention. Much research has been conducted, and numerous publications exist on the relevance of the curriculum to learners and the integration of political issues into various learning processes (Michaeli, 2014). There is an ironic inverse relationship between the importance teachers believe a topic has and their willingness to discuss it in their classrooms (Miles, 2011).

The teaching of political geography suffers from difficulties and barriers in various parts of the world (O'Reilly, 2019). For example, a study examining students' knowledge and understanding of actual geopolitical issues in the United States in the early 2000s proves a lack of attention and awareness of geopolitical issues (Holm & Farber, 2002). The researchers found that although understanding geopolitical processes was significant for teachers in times of change, the knowledge they

T. Yaar-Waisel (✉)
Oranim College of Education, Haifa, Israel

S. Wolff-Seidel
University of Wuppertal, Wuppertal, Germany

M. Hardouin, C. Leininger-Frézal (eds.), *Curricular Changes in Geography Education*, International Perspectives on Geographical Education,
https://doi.org/10.1007/978-3-031-96551-7_10

demonstrated was inferior because this article was written a year after September 11, 2001.[1] Holm and Farber found that most of the knowledge available to students is not from the education system but from the media. In 1988, there was twice as much broadcasting time for international issues as in 2000 (Holm & Farber, 2002).

A study conducted in Germany on European borders found that students need help in understanding the geopolitical implications of border demarcation, inclusion, and exclusion. The study revealed that their personal experiences significantly impact their comprehension. Interestingly, traditional textbooks and school curricula could have been more influential in helping students understand these concepts. This was illustrated by the fact that only 45% of respondents could accurately draw the map of Europe as depicted in German textbooks on a blank world map (Seidel & Budke, 2019). This is not only true for Germany: Across Europe, curricula have been reviewed, revealing that "European integration" is presented from a national viewpoint in each country, and textbooks act as "instruments of nationalism" (Sakki, 2014).

Previous studies have specifically explored the teaching of boundaries, a topic within the field of political geography (Yaar-Waisel, 2021). In this study, a teaching model for the subject was developed. The importance of domain teaching lies in two main aspects: one discusses the importance of geographical knowledge for the training of a mature person, while the other is general and raises the assumption that discussing actual controversies improves students' critical thinking skills (Sziarto, K. et al., 2014; Michaeli, 2014).

Research on the perceptions of borders in Germany has found that many people have simple and naturalistic ideas about these complex geopolitical phenomena (Seidel, 2020). In Germany, there are few references to civic education or geopolitics in the national education standards (Sitte, 2014). One possible reason for this may be the historical association of the subject with regional geography in Germany, which has been linked to ideological indoctrination, particularly during the Nazi era. This historical background may have hindered the subject from being considered a legitimate political education in schools today (Schultz, 1999; Budke, 2016).

[1] Geopolitical issues examined in this study in 2002 included: (1) International economy and international markets; (2) the power of countries; (3) cross-country cooperation and multicultural cooperation (Olympics games express the need for cooperation and internationalism); (4) environment, climate change, and sustainability; (5) human rights and basic respect for people and places are required; and (6) immigration and population growth.

10.2 The Study

Two lecturers from Oranim College of Education in Israel and Cologne University in Germany conducted the study in 2019. Both studies had the same goals, research tools, and group size. Questionnaires were developed in English but given in local languages (Hebrew and German) and then translated into English.

10.2.1 Study Objectives

Based on the study by Holm and Farber (2002), the exploratory approach aimed to initially survey the status and determine the position of prospective teachers enrolled in our institutions. Examining the existing curricula or module guides for the programs under investigation was intended to provide insights into the student survey findings within the context of the curricula.

10.2.2 Study Questions

The research guiding questions were related to the student's knowledge and understanding and the scanning of the curricula that serve as the basis for their education.

1. What level of basic geopolitical knowledge do geography undergraduate students in teacher training in Germany and Israel possess?
2. How does the curriculum of the participating students support training in the field of geopolitical knowledge?
3. What are the differences and similarities between the two samples compared to Holm and Farber's (2002)?

10.2.3 Assumptions and Research Hypotheses

Theoretical considerations and background have led to the following assumptions: Understanding geopolitical processes is crucial for creating an informed individual. This is especially important in democratic education systems, where understanding geopolitical processes should encourage civic engagement and active citizenship. Therefore, teachers have the opportunity and the responsibility to facilitate discussions and cultivate long-term interest in geopolitical issues. Consequently, the education system must address these issues and not leave them to external influences like the media. At the same time, it can be assumed that teacher training colleges are not adequately preparing teachers to address these issues. While there may be different reasons in the two countries studied, we suspect that there are identifiable parallels, leading to the following research guiding hypotheses:

Research hypotheses:

1. Curricula need to adequately prepare prospective teachers to understand, assess, evaluate, or teach within geopolitical contexts.
2. Students leave the education system without basic knowledge that allows them to understand what is happening worldwide and are unaware of the complexity of international or global processes. The students are not qualified to be "citizens of the world" and do not care about geopolitical events.

10.3 Methodology

To address the research questions, we chose a two-step exploratory approach. First, we conducted a document analysis of the study programs' curricula. Then, we followed up with a questionnaire using standardized questionnaires. The schematic structure of this research design is as follows:

10.3.1 Curricular Analysis

The analysis of German curricular documents included three primary sources:

1. "Bildungsstandards im Fach Geographie für den mittleren Schulabschluß"—educational standards in geography for the intermediate school leaving certificate, published nationally by the German Association for Geography (DGfG) in 2020.
2. The core geography curriculum at secondary schools in North Rhine-Westphalia is based, in part, on the national educational standards.
3. The module handbook for the bachelor's degree program in geography at secondary schools and comprehensive schools of the Institute for Education in Geography at the University of Cologne.

To study Israel's geography education, we examined the common textbooks published by the Center for Educational Technology (CET), known as "Kotar." These textbooks have teacher's guides, such as "Man and Environment in the Global Age" from 2013. The curriculum is published on the Ministry of Education website in its "Geography and Environmental Studies" session. We analyzed the material by looking for Keywords.

Table 10.1 Research sample (Wolff-Seidel and Yaar-Waisel)

	USA (2011)[a]	Israel	Germany
Number of students	129	41	45
How many visited other countries	88%	91%	91%
Average age	81% less than 25	24.5 years	20.5 years

[a]Data for USA: Holm and Farber (2002)

10.3.2 Questionnaires

Students were interviewed using a standardized questionnaire based on the survey instrument used by Holm and Farber (2002). The questionnaire consisted of questions about general geopolitical basic knowledge, the sizes of states, the languages spoken there, the religions mainly prevalent there, the task and importance of international organizations (e.g., UN, NATO, etc.), and about people (e.g., Vladimir Putin, Mahatma Gandhi, etc.). In this point in particular, the content of the present study differed from the survey instrument of Holm and Farber since both organizations and persons were updated and adapted to current discourses. Furthermore, a blank world map was included to test the student's knowledge of orientation and location with cities (e.g., Washington, Riyadh, etc.) and countries (e.g., Brazil, Canada, etc.).

10.3.3 Sample

A total of 86 students took part in the survey, 41 of them in Israel and 45 in Germany. The average age of the Israeli respondents was 24.5 years, and that of the German respondents was 20.5 years. One explanation for this age difference is that Israeli students usually start their studies later due to the general compulsory of military service. Regarding whether the students had undertaken trips abroad, 91% in both groups stated that they had undertaken such trips (Table 10.1).

10.4 Findings

The research findings comprise students' questionnaire responses and location data on global maps. Participants' answers form the basis of the conclusions, illuminating the primary objective of this study.

10.4.1 What Is This Place?

Questionnaires began by indicating the locations of countries on a blank map. Some countries were based on Holm and Farber (2002), and other countries were added due to geopolitical relevance, as in the media. Russia was most often correctly located of all the countries: 88.4% among German students and 81% among Israeli students. An equally large and significant country like China was situated correctly by 80% of the Israeli and 68% of the German students. Africa's third-largest country, Sudan, was correctly located by only 22% of German respondents, compared to 41% in Israel. This might result from massive emigration from Sudan via Egypt to Israel.

Nonetheless, the values in the localization area were remarkably low, considering that prospective geography teachers were surveyed (Fig. 10.1).

Only one-third of the respondents (Israel 31.5% and Germany 33.3%) knew where the capital of the USA, Washington, is in the eastern United States. Less than half of the respondents knew the capitals of Germany or Israel. Although Israelis visit Berlin, and Jerusalem often appears in news reports in Europe, only 45.8% of Israelis do. Conversely, Jerusalem is frequently mentioned in the German media, yet only 37.8% of respondents could correctly locate the city (Fig. 10.2).

The respondents were then asked to estimate the population figures of various countries. They were asked to tick size categories (note: less than 10 million, 10–50 million, 50–100 million, and over 100 million inhabitants). Significant gaps in knowledge were found. Countries with a tiny population were perceived as having a large population (e.g., Israel, Canada, and Denmark). Countries with large populations were often classified as less populated (Egypt, Turkey, or Indonesia). The population of the USA was correctly identified by 65% of Israelis, but most German students thought it was smaller (mainly 50–100 million). Less than 50% of the Israelis gave the correct size of the Iranian population. It was estimated to be more than 100 million. Potential threats may be perceived as more extensive in a geopolitical sense (Zaga & Yaar-Waisel, 2023).

The next question referred to the most widely spoken languages in certain countries based on the answers in free-text fields. The answers were largely correct. Most students knew the languages spoken in Morocco, Brazil, and Chile. Both groups stated at the beginning of the questionnaire that they had already traveled abroad. Israel is not associated with Hebrew, with only 28.9% correct answers from the German students.

Finally, the question about the most widespread religion in different countries was asked. In the case of the Israeli respondents, well over 50% of the respondents could give correct answers. The German students scored weaker overall; this is especially noteworthy for Israel, where only 51.1% of the respondents gave the correct answer "Jewish." There appears to be a lack of knowledge about Israel, which is surprising considering German history and its coverage in school lessons.

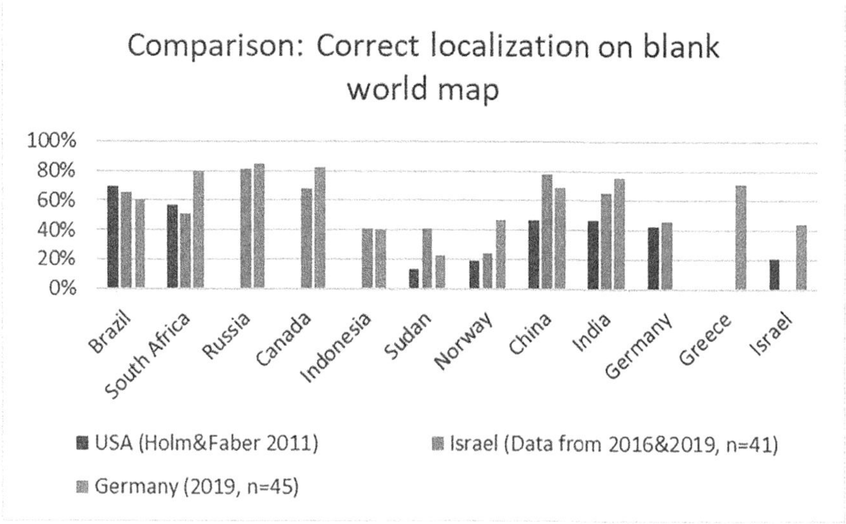

Fig. 10.1 Research findings: Correct localization on a blank world map (Wolff-Seidel and Yaar-Waisel)

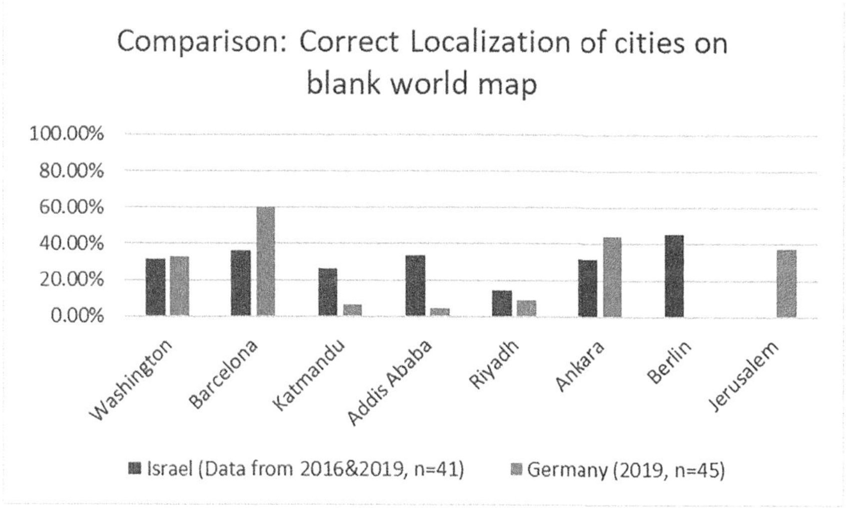

Fig. 10.2 Research findings: Correct localization of cities on a blank world map

10.4.2 Names in the News

Following Holm and Farber (2002), the task was to correctly identify people and organizations that frequently appear and are named in the media (Fig. 10.3).

It was interesting, for example, that in the case of people, 95.6% of Germans and 77.3% of Israeli students correctly identified Vladimir Putin as the President of the Russian Federation since Putin has appeared constantly in the German media since the Ukraine conflict in 2014 at the latest. At the same time, he was often mentioned following the Syrian civil war. Interestingly, only 13% of the US respondents in Holm and Farber's study correctly identified the Russian president, probably as it occurred 18 years earlier. King Abdullah of Jordan, head of state of a direct neighbor of Israel, was correctly identified by 74.8% of the respondents from Israel. The German students were only able to locate him by 17.8% correctly. This is one example of the influence of distance on geography knowledge, which will be discussed later.

Even more remarkable than these examples of people was the level of knowledge of the students interviewed concerning international organizations. Only 29.6% of the German respondents could identify NATO as a North Atlantic Treaty Organization or at least as a "Western defense alliance." Since Germany is a member and participates in NATO missions at the time of the survey, this is a considerable deficit for almost 70% of the respondents. On the other hand, Israeli students were virtually unfamiliar with NATO, with only 4.8% of respondents correctly identifying the organization. By comparison, 43% of respondents in the American study were able to locate the organization correctly. The Taliban, as the former and now active government of Afghanistan, was also not very well known everywhere (Germany: 62.2% and Israel: 26.3%).

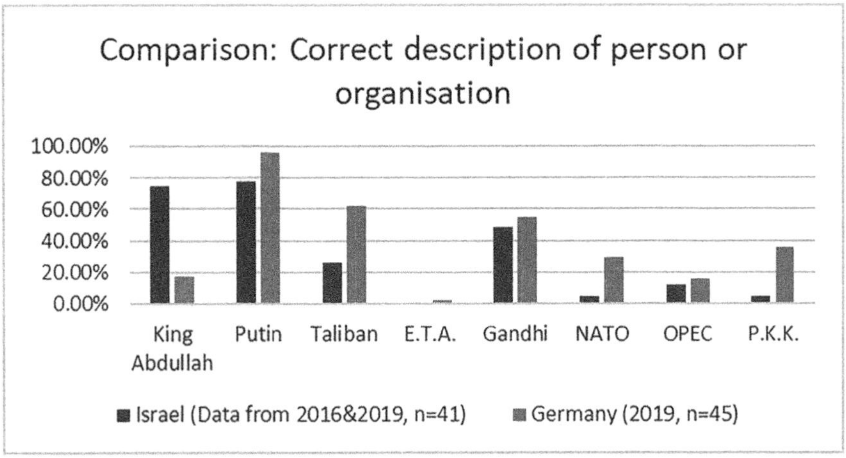

Fig. 10.3 Research findings: Correct description of person or organization (Wolff-Seidel and Yaar-Waisel)

In terms of organizations, the results of the following question were quite remarkable. According to Holm and Farber, only 6% of the respondents had correctly named the number of UN member states. In our survey, only 14.3% of the Israeli and 13.3% of the German assessments were almost correct or deviated only by a few values from the actual number between 190 and 200. Nearly all others estimated the numbers at least 50 states higher or lower. The frequent mention of 15 suggested that the UN and the UN Security Council were confused here. Overall, it is remarkable that there was so little knowledge about the United Nations.

10.4.3 Participants' Responses

After completing the questionnaire, the students were asked to rate it. The results indicated that they were disappointed and frustrated with their shortcomings but also highly motivated. The results were analyzed and are presented in Table 10.2.

The students have shown a keen interest in learning and understanding the topics. This underscores the importance of including geopolitical issues in school curricula and teacher training programs, particularly for geography teachers. According to the students, the subject examined in this article is highly valuable.

Table 10.2 Participants' responses (Wolff-Seidel and Yaar-Waisel)

Category	Israel	Germany
Reflection	"I realize how much I don't know." "Disappointing, I don't know enough," "I have no idea of anything going on around me," "My familiarity with the world is very low," "Very confusing. Made me want to learn more", "Made me want to learn, I realized I didn't know," "Made me embarrassed, these are things I need to know," "Questions I never encountered", "Reflective, basic, and important."	"Find interesting how little one knows," "Shows your own gaps in knowledge," "A lot of questions. Exciting also to discover the gaps in one's own knowledge", "Good, because I realize I don't know a lot of things yet," "I think it's very important for general knowledge." "Should be discussed much more intensively in the seminar since it belongs to the general knowledge," "Shows you certain limits."
Positive feedback	"Very interesting!" "Refreshing and cool," "Creative questions," "The questionnaire is excellent; I checked it myself."	"A very good review of general education," "Good as an introduction to the seminar," "Find it very diversified."
Other	"Challenging"	"General knowledge!"

10.5 Discussion: Existing Curriculums and Textbooks

German and Israeli curricula were studied for this research, and textbooks and online teaching materials were reviewed. The content from each country will be presented in separate sections.

10.5.1 Geopolitical Terms in the Existing Geography Curriculum and the Geography Textbook in Israel

A simple search of geopolitical terms in the Israeli Ministry of Education's Geography website shows that almost none of the geopolitical field's main terms exist. Words such as "Minorities," "Autonomy," "Political conflict," or the "European Union" have not been represented, neither in the Middle school curricula nor high schools—nevertheless, "Geopolitics" as well.

Two terms are excluded from this situation and are being represented in the curricula: Globalization and Borders. Globalization is a topic that has been thought about since the first year of middle school (seventh grade). The title of the subject is: "Globalization amplifies changes in the world," this chapter includes many terms: "The global village," "Technology," "Open market," "High tech industry," "import," "export," "developed country," "undeveloped country," "Americanization," "consumer culture," "Glocal," and "develop index." This list makes us easily understand that globalization learning is based on its economic commentary. Most of the dilemmas discussed have a financial concern.

The most common geopolitical subject that appears in the geography curriculum is borders. This topic exists either in the middle or high school curricula. It might not be surprising for a state with 1000 km of borders but only 22,000 sq. km. As published before (Yaar-Waisel, 2018), much of the content focuses on the subject's historical significance: How and when were Israel's boundaries established? Political dilemmas are rarely addressed in the middle school curriculum but are included in the high school curriculum. Additionally, only two study hours are recommended to teach the subject of borders in middle schools.

It should be mentioned that Middle Eastern studies are included in geography curricula, but there are almost no learning hours for this topic in middle schools and more in high schools. Only very few students choose geography as their chosen subject in high school (Zaga & Yaar-Waisel, 2023).

In the geography curriculum, the "United Nations" is discussed in the context of globalization and its connection to the global issue of the climate crisis. However, this context needs to mention the European Union and other organizations, such as Organization of the Petroleum Exporting Countries (OPEC).

Immigration was one of the subjects learned during the Syrian Civil War 2005–2006. Usually, immigration has been understood as a history lesson for Jews during the last 150 years.

All geography textbooks are published by the CET (2013), a private publishing house called Kotar.

The seventh-grade textbook, "Man and Environment in the Global Age," includes most geopolitical terms. It was upgraded in 2013.

Indeed, this textbook extensively includes the following geopolitical topics:

1. Globalization. An Economic Process. Widely explained. Include many economic terms about the cultural aspects of globalization, such as Americanization and the disappearance of local cultures[2]
2. Globalization in the context of environmental issues. "Environmental justice"
3. World's population, minority groups. Spatial inequality in different aspects (the development index)
4. Immigration and refugees—widely explained in global aspects

It provides a comprehensive discussion of historical and current migration factors, detailing various immigration routes from Ireland in the past to more recent immigration from Syria and other countries.

Developed and less developed countries: How does migration affect them? What is the extent of its impact?

5. Global organizations are studied: the United Nations and the European Union

The textbook offers a wide range of teaching methods for the chapter, including pictures, maps, quantitative data, a link to videos, and more.

While the term "geopolitics" is not explicitly mentioned in the book, it is essential to highlight that the content aims to encourage students to think critically and evaluate available information. This involves identifying the advantages and disadvantages and understanding the benefits and drawbacks of geopolitical processes. When school principals choose to teach geography in the seventh grade, this book is supposed to be studied by all Israeli students.

A teacher's guidebook accompanies this book. This guidebook includes aspects of geography and didactic aspects and addresses the value aspects of teaching the book.

Based on the information discussed, it can be concluded that Israeli students are expected to be familiar with two important topics within the realm of political geography: globalization and borders. The textbook facilitates comprehensive learning about globalization, while the teacher guide allows educators to teach the subject in a diverse and current manner. Instructional planning demonstrates professional consideration, reflecting the challenges of teaching in the twenty-first century. Furthermore, an emphasis on the value aspect of teaching geopolitics is incorporated, which holds significant importance.

[2] "Globalization is a collective name for the processes of economic, social, cultural and political connections that humans have forged with individuals and with communities all over the planet." https://pop.education.gov.il/tchumey_daat/geography_adam_sviva/chativat-beynayim/noseem_nilmadim/globalizatia/

Not all students study geography in seventh grade; for most, it marks the end of their global geography studies until they finish school. From eighth grade until high school graduation (ages 14–18 years), students are not required to learn about geopolitical issues in a global context. As previously mentioned, borders are extensively covered, but only in the context of Israel's borders, particularly from a historical perspective.

10.5.2 Geopolitical Terms in the Existing Geography Curriculum in Germany

In Germany, each federal state has its own educational curricula due to the federal structure. Still, there are also national "Educational Standards in Geography for the Secondary School Certificate" established by the DGfG in 2020. These national standards are used as a reference in developing the curricula at the state level. Teachers refer to the core curriculum for geography at Haupt-, Real-, and Gesamtschulen in North Rhine-Westphalia, the national educational standards of the DGfG, and a module handbook that reflects the study contents of the students surveyed.

The term "geopolitics" was not found in the documents we examined. However, there are passages about political conflicts in the national educational standards. Specifically, three out of the six terms explicitly mention the knowledge about conflicts that should be imparted to the pupils. This falls under the "Subject Knowledge" competence field, where students can systematically explain the effects of space use and design (such as migration and resource conflicts) using specific examples. A second finding of the term migration coincides with the geopolitically relevant term war: "Students can explain the real consequences of social and political constructions of space (e.g., wars, migration ...)" (There, 15). Interestingly, the word "migration," which is also examined, is mentioned here at the same time. Moreover, both terms appear in the competence area "assessment/evaluation" in this formulation: "Pupils can apply geographical knowledge and the criteria mentioned above to assess selected geographically relevant facts, events, challenges and risks (e.g., migration, ..., conflicts when cultures meet, civil wars, resource conflicts)" (cf. ibid., 24). Although the national educational standards address the thematization of geopolitical conflicts and wars, this is reflected differently in the more stringent core curricula of the federal states. The concept of conflict is addressed in eight instances in this curriculum. It mainly refers to "space use conflicts," which are disputes related to questions like "Should this forest be cut down for resource extraction?" (MBS, 2020, 12). This becomes clear, for example, in the following quotation: "The students explain selected aspects of the conflict of goals between economy and ecology in tourism regions" (There, 19). The curriculum examined did not include terms such as politics, war, conflict, or geopolitics, which apply to the student teachers surveyed. The term "migration" is only found in the description of an intermediate seminar on environment and society. "*In environment and society, physical-geographical and human-geographical questions and knowledge are deepened on selected topics such* as migration" (Institute for Geography Education Cologne, 2015, 12). Terms such as "minorities," "autonomy," or "European Union" did not appear.

The geopolitical concept of the border, which is central in our view, needs to be addressed in this sense. The term appears in five educational standards (DGfG, 2020, 5; 16; 42; 44). In three cases, physical–geographical border concepts are meant. In another case, it is about "*scientific knowledge acquisition and its limits*" (There, 5). This is also related to the fact that the term "Grenze" is used for borders, boundaries, maximums, and limits in the same form in German. In the last case, the competence of the pupils is indeed demanded here in "specialized knowledge," but only as an objective: "*Pupils can mark spaces according to certain characteristics and delimit them comparatively*" (There, 16). In this respect, no reflection on boundary demarcation processes is required here, but only the simple demarcation itself. The term "border" was not found in the other two documents examined.

10.6 Summary and Conclusion

Despite its small size in terms of territory and population, Israel faces numerous critical geopolitical issues. Geopolitical events near and far significantly impact the daily lives of students and teachers. Researchers had assumed that it was impossible to avoid geopolitical issues, but their findings proved otherwise. The research revealed that students need to be better informed about geopolitical issues, particularly those related to the Middle East, including its leaders and organizations.

Germany is significantly larger in size and population than many other countries. It plays a crucial role in geopolitical events both within Europe and globally. In many ways, its geopolitical situation resembles that of Israel. It is essential to recognize that geopolitics cannot be ignored, yet students' understanding of this area is often limited.

The survey results have shown that students already have deficits in basal factual knowledge in many areas. Although they often perform somewhat better overall than the American study (Holm & Farber, 2002), these deficits are sometimes considerable. Against the backdrop of the research questions, it can be assumed that these students have difficulties classifying, assessing, or evaluating geopolitical events and certainly do not communicate with them in teaching contexts. The analysis of the curricular documents also shows that, although the national educational standards certainly demand that geopolitical issues be addressed, this needs to be implemented in the more binding and relevant core curricula or the students' curriculum.

It was found that differences in geopolitical reality between Germany and Israel could be more effective in teaching geopolitical issues, neither in the studying hours given in the curricula, teaching methods, or the scope of the material asked to be taught.

This study took place when Russia entered Crimea and the Syrian civil war erupted. Apart from acknowledging Putin, it did not yield a significant geopolitical impact. However, students value geopolitical knowledge, considering it both interesting and vital.

Teaching materials should be updated regularly, and educators should be trained to address questions that encourage critical thinking, even if they involve emotional and political aspects. As mentioned earlier, the educational system often overlooks this crucial issue.

Teacher training should prioritize the importance of teaching political subjects and equip teachers with the skills to overcome teaching challenges. Student-teachers, or teachers-in-training, should be familiarized with teaching these subjects during their training period. They should have opportunities to practice teaching through classroom discussions and develop the necessary skills for facilitating such discussions. There is a crucial need to expose students to diverse opinions without imposing a one-sided stance or causing confrontation, which could be poorly received by students, parents, or the education system. Geography teachers should be aware of disparities between the current political realities and the maps used in class, including those found in atlases and textbooks.

Teachers should prioritize educating young people in political thinking, evaluating alternatives, and drawing conclusions. In a democratic society, students need to develop the skills for independent thinking and formulate their positions based on historical and political reality. When students are exposed to different viewpoints and diverse opinions, they can more effectively and objectively assess the discussed issues.

This study revealed that teachers avoid teaching geopolitical issues because they lack sufficient knowledge and preparation. Let us work on changing this!

References

Budke, A. (2016). Potentiale der Politischen Bildung im Geographieunterricht. In A. Budke & M. Kuckuck (Eds.), *Politische Bildung im Geographieunterricht* (pp. 11–23). Steiner.

Holm, G., & Farber, P. (2002). Teaching in the dark: The geopolitical knowledge and global awareness of the next generation of American teachers. *International Studies in Sociology of Education, 12*(2), 129–144. https://doi.org/10.1080/09620210200200087

Michaeli, N. (Ed.). (2014). *Ken beveit sifrenu: Ma'amrim 'al hinuch polity [political education: an anthology]*. Hakibbutz Hameuchad.

Miles, W. F. (2011). Border pedagogy in Israel. *The Middle East Journal, 65*(2), 253–277. https://doi.org/10.3751/65.2.14

O'Reilly, G. (2019). *Aligning geopolitics, humanitarian action and geography in times of conflict [Textbook]*. https://doi.org/10.1007/978-3-030-11398-8

Sakki, I. (2014). Social representations of European integration as narrated by school textbooks in five European nations. *International Journal of Intercultural Relations, 43*, 35–47. https://doi.org/10.1016/j.ijintrel.2014.08.010

Schultz, H. D. (1999). Geographieunterricht und Gesellschaft. Kontinuitäten und Variationen am Beispiel der klassischen Länderkunde. In H. Köck (Ed.), *Geographieunterricht und Gesellschaft* (pp. 35–47). Selbstverlag.

Seidel, S. (2020). *Grenzen im Kopf - Untersuchungen von Lernvoraussetzungen und digitalen Spielen zur Thematisierung räumlicher Grenzen im Geographieunterricht* [Dissertation work, Universität zu Köln].

Seidel, S., & Budke, A. (2019). "A border is a ban" – Students' conceptual understanding and experiences of Europe's borders and boundaries. *Review of International Geographical Education Online (RIGEO), 9*(1), 82–101. https://doi.org/10.33403/rigeo.573476

Sitte, C. (2014). Politische Bildung im Geographieunterricht. *Geographie aktuell & Schule, 36*(208), 29–35.

Sziarto, K. M., McCarthy, L., & Padilla, N. L. (2014). Teaching critical thinking in world regional geography through stakeholder debate. *Journal of Geography in Higher Education, 38*(4), 557–570. https://doi.org/10.1080/03098265.2014.958658

Yaar-Waisel, T. (2021). Teaching borders: A model arising from Israeli geography education. *Borders in Globalization Review, 2*(2), 52–60. https://doi.org/10.18357/bigr22202119633

Yaar-Waisel, T. (2018). Bordering on the impossible: Optimistic planning of border regions. *Journal of Borderlands Studies, 36*(3), 389–404. https://doi.org/10.1080/08865655.2018.1531052

Zaga, M., & Yaar-Waisel, T. (2023). Mapping the Middle East: Israeli student's geographical knowledge and perceptions. *Journal of Borderlands Studies*, 1–24. https://doi.org/10.1080/08865655.2023.2255191

Tal Yaar-Waisel is a Senior Lecturer at Oranim College of Education in Israel. She specializes in political geography, border studies, and teaching skills. She is a lecturer in a training teachers' program and teaches geography education (Didactics). Some of her previous research (Google Scholar—Yaar-Waisel) focused on international collaborations in teaching and learning. Other studies included geopolitics and environmental planning of border regions in Europe and Israel.

Sebastian Wolff-Seidel is a lecturer at the University of Wuppertal, focusing on geography education, political literacy, and digital media. His current research explores digital games in education, spatial boundaries, and sustainability, emphasizing education for sustainable development (ESD) and open educational resources (OER).

Chapter 11
Teaching the Concept of Territory in a Bachelor of Technology in France: Adjustments to Deal with a Bureaucratic Reform

Régis Keerle

11.1 Introduction

This text presents a case study of how a geographer teaches students about territory in an *Animation sociale et socioculturelle* (ASSC) curriculum in France. This teaching cannot keep a close culture of disciplinary, but neither is it the subject of a clear standard. The reform of this curriculum, implemented in 2021, sheds light on the problematic nature of the development of this standard. It depends on the relationships between a large number of agents whose power relationships are highly asymmetrical, but specific to higher education insofar as the mechanisms for monitoring training, even, in this case, for professional purposes, remain relatively weak. As a result, the geographer's place in this curricular chain is not established.

The first part of this text presents the context in which the reform of the Bachelor's degree applies to training in ASSC at the *Instituts Universitaires de Technologie (IUT)*. Its second part presents my understanding of teaching students about territory in this training, and the adjustments, first the one envisaged but forsaken and then the one finally introduced, to apply this reform. Excerpts from students' work are used to spell out their understanding of my teaching goals.

R. Keerle (✉)
UMR 6590 ESO Espaces et sociétés, Université Rennes 2, IUT de Rennes, Rennes, France
e-mail: regis.keerle@univ-rennes2.fr

M. Hardouin, C. Leininger-Frézal (eds.), *Curricular Changes in Geography Education*, International Perspectives on Geographical Education,
https://doi.org/10.1007/978-3-031-96551-7_11

11.2 Background to the Teaching of the Concept of Territory in Social Work Training in France

11.2.1 *Animation sociale et socioculturelle in the Bachelor's Degree Reform*

Animation sociale et socioculturelle (*ASSC*) "… can be defined as a process of distribution in various populations of the capacity of sensitive experience, expression, critical thinking, initiative and creativity" (Lafortune et al., 2010, 2). This field of activity is represented in all states in the Western world or which accept its influence, but in different forms. According to the above authors, in English-speaking countries, ASSC corresponds more or less to community education, continuing education, and learning society. Outside France, these different areas of activity are encompassed within the notion of social work, and social work training is often more closely linked to university education (Chauvière & Gaillard, 2020).

In France, *ASSC* is not officially clearly included in social work, which is most often defined in relation to diplomas designed by the Ministry of Social Affairs, whereas the first youth worker diplomas were designed by the Ministry of Sport and Youth (Lebon, 2020). However, specialists in the analysis of social work most often include *ASSC*. One of the common features of all these areas of activity is the wide range of training courses, both within and outside the university (*Établissements de Formation au Travail Social—EFTS*; Bergougnan & Fondeville, 2021). Within the university, these training courses are provided in particular in the *Carrières sociales* branch of the *IUT*. Teachers are in the minority (one-fifth of representatives) on the *Commissions pédagogiques nationales* that establish the curricula for *IUT* branches, as well as on the *Commission consultative nationale* (2 members out of 17) that assesses training courses at the level of each *IUT* "... according to criteria largely influenced by the standards applicable to *EFTS*" (Chauvière & Gaillard, 2020, 90).

As part of the harmonization of higher education curricula in the European Union (the Bologna process which began in 1999; Grek, 2024), the *Diplôme Universitaire de Technologie* (*DUT*; level 5 of the European Qualifications Framework) is replaced, since 2021, by the University Bachelor's degree in Technology (UBT; level 6 of the European Qualifications Framework). The Bachelor's syllabus in the *Carrières sociales* branches comprise five «skills.» Two of these (see Sect. 11.2.2) are common to all five branches (ASSC, *Assistance sociale, Coordination et gestion des établissements et services sanitaires et sociaux, Éducation spécialisée, Villes et territoires durables*).

These "skills" are specified into "essential components," which are practiced in "professional situations," thanks to "critical learning," used at increasing skill levels between the first and third years of the UBT. Students acquire this critical learning through "Resources" (R: teaching modules), "Learning and Assessment Situations" (LAS; tasks responding to a problem encountered in the workplace), and internships.

There is almost no reference to geography in the *Carrières sociales* Bachelor's syllabus (*Ministère de l'Enseignement supérieur, de la Recherche et de l'Innovation*,

2022), as the competency-based approach is supposed, as part of the Bachelor's degree reform, to "... encourage multi-disciplinary dynamics around technology within teaching teams"[1] (Georges & Poumay, 2020). This reform, boosted by *IUT*s' Board of directors (Dansac, in Boucher et al., 2024), is highly ambitious, as "the guiding principle in training engineering is no longer based on the transmission of knowledge and its practical application, but on a systemic model in which knowledge would certainly be assessed for its own sake, but above all in terms of its use and employ in professional environments and situations" (*Groupe de Travail Commission Pédagogique Nationale/Commission Consultative Nationale des Instituts Universitaires de Technologie*, 2020, 3). It is being introduced at breakneck speed, between the Order of the *Ministère de l'enseignement supérieur, de la recherche et de l'innovation* (December 6, 2019) reforming the vocational bachelor's degree and the start of the 2021 academic year.

"With regard to the timing of the planned changes, the *Assemblée des Chefs de Départements Carrières sociales* (*ACD Cs*)[2] points out that establishing a coherent national curriculum is a process that takes time. The reforms to the National Curriculum involved work that lasted at least a year and a half, even though the frameworks were known and the duration maintained. The move to 3 years poses major challenges in terms of reflection on progress. It is therefore concerned about the timeframes envisaged for their implementation.

Lastly, the *ACD Cs* notes that it was not involved in the discussions on this reform, and that the process put in place, which is very top-down, takes little account of the specific features of its various branches. It regrets the lack of consultation and collegiality upstream of this reform ..." (*Motion de l'Assemblée des Chefs de Départements Carrières sociale*s, November 19, 2019).

The *Laboratoire de Soutien aux Synergies Education-Technologie* (LabSET) at the University of Liège, official support for the reform at the *IUT*s' Board of directors, was not bothered by the remarks made in 2020 by representatives of the *Commission pédagogique nationale Carrières sociales*, and those of the teachers' working groups of the five future curricula.

11.2.2 Teaching the Concept of Territory in the Bachelor's Reform

The Bachelor's syllabus of the *Carrières sociales* branches provides guidance on the place given to the geographical approach to the notion of territory. Both skills common to its five branches mention it in their essential components, respectively:

[1] Of the 24 specialities offered by the IUTs, 16 correspond to purely industrial activities.

[2] These chiefs are most offently teachers, elected for 1–3 years, more or less on a voluntary basis, or appointed by the IUT directors.

- For Skill 1, “Designing interventions adapted to the challenges facing society/By developing a well-founded point of view on issues relating to social intervention and territorial cohesion” (*Ministère de l’Enseignement supérieur, de la Recherche et de l’Innovation*, 2022)
- For Skill 2, “Building partnership dynamics/By mastering the institutional framework and the dynamics of agents at all territorial levels” (idem)

For Skill 1, the critical learning involved consists of “identifying elements of context (social, economic, cultural, demographic, political, territorial, educational, and so on) and questioning the agents involved.” Although the emphasis is on “territorial cohesion,” the Bachelor’s syllabus suggests that a geographical conception of the territory has its place in it. For Skill 2, the critical learning involved consists of “mobilising partners and diversifying contacts (geographical scope, inclusion of all publics, diversity of areas of expertise . . .).” This time, the oxymoron “geographical scope” suggests that the notion of territory is understood in its purely political sense.[3] For these two skills, common to all five UBT syllabi, the notion of territory is only mentioned in critical learning at their first skill level (first year of the UBT syllabus). It is only mentioned again in the content of a Resource at the second level of UBT skill 2.

For the specific skills of the *ASSC* branch, the place of the notion of territory is used differently. It appears in one of the essential components of Skill 3 (“Strengthening the public’ individual and collective capacity to act/By taking into account the specific characteristics of the territory and the contexts of intervention”), but only at this part of its architecture. It is also mentioned for Skill 4 (“Implementation of educational approaches and facilitation techniques as part of a project-based approach”), but only as part of the critical learning addressed at this third skill level (third year of the UBT). This is also the case for Skill 5 (“Contribution to the development of the professional field of *Animation sociale et socioculturelle*”). In the case of these two skills, the wording of the critical learning outcomes concerned (“Co-ordinate interventions on the scale of structural and territorial projects” and “Implement interventions aimed at the empowerment of individuals and groups and the development of territorial resources”) leaves some doubt as to the place of geography in the meaning of the notion of territory. However, a description of one of the Resources in the second level of Skill 5 (“The comparative dimension may be addressed in a more territorial approach, on an international or interregional scale, between urban and rural realities, and so on”) removes this doubt, at least for Skill 5.

[3] It would have been better to use the expression “territorial levels” rather than “territorial scales.”

11.3 From "Territorial Diagnosis" to the LAS "Introduction to Territorial Survey and/or Diagnosis"

11.3.1 "Territorial Analysis" Rather Than "Territorial Diagnosis"

My experience of teaching in successive *ASSC DUT* syllabus since 2005, a field that is still scientifically relatively unexplored (Camus & Lebon, 2015), has convinced me that the notion of territory remains a buzzword (Giraut, 2008) that the development of case studies (Greffier, 2013) or systems-inspired works (Moine & Sorita, 2015) has not yet succeeded in establishing as a topos (Thémines, 2016). Some students are aware of this situation:

> I did my second internship at the University Hospital Center Grenoble Alpes and more precisely in the rehabilitation institute located … in Echirolles ... In the first semester of the UBT, in one of the resources, we had seen that it was interesting to be able to have an interview with an elected official of the municipality to have information about the territory, the action of the internship site... I think that in this case an interview with an elected official of the municipality would not have been relevant. An elected official of which municipality? Questions about which territory? However, an interview with other agents would have been more relevant and would have probably been complementary to the one already carried out with the professional tutor. Indeed, an interview with the director or even a senior manager of the internship site as well as another one with a caregiver or a secretary would have allowed to compare different points of view on the same issues (Cadoret, 2022).

Until 2020, my teaching as a geographer amounted to 48 h per student (at three tutorials; with regard to 1800 h of training for the 3 years of the BUT program), the rest of my teaching time (192 h in total) being used for other cross-disciplinary teaching and student support specific to *IUT*s (small groups tutoring). It is hardly possible, with such a volume of teaching hours that can be devoted, either directly (Resource: Territorial analysis) or indirectly (Resources: Population dynamics and Cartography), to claim to cover, even quickly, all the questions raised by the question of the ontology of the notion of territory and its relationship with other (more or less) fundamental notions in my training discipline. I, therefore, chose the objectives that would be easiest for the students to achieve and, as objective as possible, given the diversity of the professional situations they encountered during their internships.[4] These objectives are as follows:

- In R Territorial analysis:[5] to present the territorial contexts that are relevant for understanding the actions carried out in the two internship sites and to present the territorial contexts of the agents involved in these actions

[4]These internships are carried out in sites as various as: *Association, Centre social, Établissement d'Hébergement pour Personnes Âgées Dépendantes, Établissements médico-sociaux, Maison des Jeunes et de la Culture…*

[5]And not "territorial diagnosis," an appellation with an exhaustive dimension, in my opinion beyond the reach of students who can only devote a limited amount of time to this work, one of their 37 different works assessed in the first year of the UBT.

- In R Cartography: to understand how a map is designed (including unresolved problems—projection, scale, framing, performativity of the map's content—as well as their workarounds) and its uses (simple direction aid or areal description tool), and experiment with the notion of a mental map
- In R Population dynamics: to illustrate the diversity of the reasons for territorial anchoring in relation to different themes covered in tutorials (one theme per session: heritage, migration, culture, micro-location, mastering of the use)

These courses have been reduced to 36 h per student as part of their adjustment to the new UBT syllabus.

11.3.2 An Attempt to Specify the Teaching of the Concept of Territory into Skill Levels

The *Carrières sociales* department of the Rennes' *IUT* was necessarily a weak agent in the consultation within the five teachers' working groups corresponding to the five branches of the *Carrières sociales* UBT, as it offered only one of them (*ASSC*). The possibility of local adjustment of the National curriculum up to 40% per year probably deterred some *Carrières sociales* departments (such as the Bordeaux one, where most of the geographers teaching in these departments are based) from challenging the dominant guidance of the writing of the Resources' and LASs' contents, very much in line with the *LabSET* recommendations. In support of this hypothesis, in the *Carrières sociales* department of Rennes' *IUT*, only 4 colleagues (among 17) participated, more or less occasionally, in the *ASSC* teachers' working group, apart from the successive heads of the department. In addition, within the *Carrières sociales* department, a rift developed between colleagues who wanted to involve temporary workers and former students in the work of writing these contents, which did not facilitate the emergence of alternative proposals to the *LabSET* standard.

Given these limits to the dialogue concerning the writing of this content, on December 24, 2020, I first proposed to my colleagues in the *Carrières sociales* department of the Rennes' *IUT* a framework for writing the content of a LAS for the first skill. Entitled "Territoriality, territories, territorialization," each of these terms referring specifically to each of its three skill levels, this LAS aimed to raise students' awareness of the notion of territorial anchoring of social actors (Debarbieux, 2014), the political dimension of territories, and the processual nature of territorialization, respectively. As none of my colleagues responded to my proposal, I did not seek to advocate this LAS to the *ASSC* teacher's working group, as I did not have a mandate to do so. Nonetheless, I maintain its relevance, with a view to the future development of training (Tables 11.1, 11.2 and 11.3):

Table 11.1 Proposal based on an imposed framework: LAS UBT first level

Imposed framework	Proposal
Title of the LAS	*Territoriality, (territories), (territorialization)*
Targeted skill (initial level)[a]	Strengthen the public' individual and collective capacity to act
Targeted learning outcomes	Understand the territorial dimension of the facilitator and ASSC
Description of the LAS in a few lines	The future professional must identify, within the framework of the internship structures, the institutions, policies, and mobilizations that are used as holds or nets by the agents (holds for the players in the territory, nets for the stakeholders of the territory)
What professional problem does it propose?	The future professional must consider the effect of the facilitator's action in relation to the institutions, policies, and mobilizations that are used as holds or nets by the agents (holds for players in the territory, nets for stakeholders of the territory)
What are the pedagogical forms (case study, simulated activity, context of an internship, a tutored project, etc.)?	Individual work in the context of an internship
How can the course be spread out over an academic year? Which possibilities of decline in two semesters?	Spread over the academic year
What is the volume of work for the student (number of hours)? What are the supervision needs (number of hours)?	Framing 48 h
What are the resources that the student will have to use and combine (in connection with existing or to-be-found resources)?	Documentary research techniques, observation, geography, and translation into spatial dimension of all other available resources
What devices and supports will enable self-evaluation and self-questioning?	Feedback on the corrected assessment
What feedback and support methods will be offered to the student in a formative dimension throughout the LAS?	Answering questions at the office
What assessment procedures need to be designed to ensure that this LAS contributes to the achievement of the target competency level? List of criteria and indicators (linked to the essential components) Nature of deliverable Feedback Remediation	– Evaluation criteria: relevance of the information collected – Deliverable: file – Feedback, catching up: in UBT 2A
How does this LAS integrate a portfolio approach?	Diary for monitoring the implementation of the file and its effects

[a]Due to reform' haste, only one skill could be targeted at the first level of the LAS

Table 11.2 Proposal based on an imposed framework: LAS UBT second level (Excerpt)[a]

Imposed framework	Proposal
Title of the LAS	*(Territoriality), territories, (territorialization)*
Other targeted skills	All other four skills
Description of the LAS in a few lines	It is a question of identifying the various ways in which the facilitator can or cannot "disrupt" the institutions, policies, and mobilizations that contribute to the construction of territorial identities

[a]As there is no major change in some lines of the LAS' second and third table layout, only two excerpts have been retained

Table 11.3 Proposal based on an imposed framework: LAS UBT third level (Excerpt)

Imposed framework	Proposal
Title of the LAS	*(Territoriality), (territories), territorialization*
Other targeted skills	All other four skills
Description of the LAS in a few lines	It is a question of identifying the various figures of territorialization proposed by the facilitator to the institutions, policies, or mobilizations that contribute to the construction of territorial identities.

11.3.3 A Withdrawal on the First Level of Skills

Following this failure, and in anticipation of the introduction of the first year of the UBT at the start of the 2021 academic year, I have, therefore, devised the writing of half of the LAS "Introduction to Territorial Survey and/or Diagnosis"[6] content in June 2021. The students' work will be assessed on the basis of one or more examples of the implementation of types of spatial ideology from their internship experience or from other LASs. The first of the four tutorials focused to this part of the LAS is focused to the analysis of a text on the notion of spatial ideology (Gilbert, 1985). The first example below shows that students are able to demonstrate their understanding of the concept:

> I ... will present you the Oriental Park of Maulévrier, the biggest Japanese garden in Europe. During my internship … I went to the *Établissement d'Hébergement pour Personnes Âgées Dépendantes* of Maulévrier during one week.

- *Localist and Universalist Ideologies*

The localist ideology is articulated within the oriental park on several points: It becomes a place apart, outside the city, which can be categorized as an enclosed space despite the comings and goings of the public. This place proposes a pause in

[6]The other half of the LAS, presented and assessed by a colleague with extensive professional experience in the animation field, requires students to analyze the usefulness of the tools used for their territorial "diagnosis" and to reflect on the construction of these tools.

time, and it erases all that surrounds it. The Japanese garden makes sense through its social and functional characterization in a localist ideology. Indeed, in addition to offering a moment of relaxation to visitors, it offers them many other services: knowledge of Japanese culture, restaurants, and hotels nearby. It creates a territorial link within Maulévrier by creating a territorial and social proximity, conducive to encounters and discovery. The garden plays on its action in the daily fabric of encounters: welcoming more than 216,000 visitors during the year 2020.

The universalist ideology of the oriental park is very present. Indeed, being considered as the largest Japanese gardens in Europe, it has a very strong worldwide visibility. Thus, being part of a global dynamic, it opens itself to the world and propels Maulévrier into the same global dynamic.

- *Ideology of Identity*

We could consider that the Oriental Park of Maulévrier does not have an ideology of identity because it is part of a culture other than the one present in Maulévrier: the Japanese culture and not the French culture. However, it is part of a dynamic of its own, thus separating it from the city of Maulévrier and the identity of its territory.

- *Ideology of Integration*

The Japanese garden keeps a very strong link with the city of Maulévrier; it cannot be detached from it. Indeed, the management of this space is carried out according to the management of the other spaces of the city.

The oriental park is above all integrated into a territory before being a space identified as a proper place: in a city, a conurbation, a department, a region, a country, a continent, and a world. It is, therefore, subject to the polarization of these: if this park were located elsewhere, for example, in Africa, it would not have the same territorial impact on the world but also on its immediate environment.

- *Ideology of Complementarity*

This uniform space that remains to be developed is in perpetual change and evolution. However, it has one primary function: it is a cultural space. It is nothing else than that, this garden has been arranged in order to create a place of culture and leisure. However, it generates many stakes since it is the use that one makes of it on a collective agreement and is thus implanted on a larger territory than that of the city of Maulévrier.

Being a neutral space that could be considered out of time and out of the city and society, the Japanese garden of Maulévrier proposes an ideal of cooperation and understanding between all, between the managers of the site and the visitors, between the young and the old, between women and men, and between the rich and the poor despite the fact that the entrance to the park is not free. Thus, it becomes a place of gathering, conducive to exchanges, discoveries, discussions, and good understanding, erasing tensions for a moment (Piou, 2022).

The second example shows that students are able to inspire hypotheses that make it possible to understand the lack of relevance of the notion of territory for many professionals in the field of ASSC:

During my internship, I attended a team meeting in which a particular subject caught my attention. The director wanted to defend his vision of *Animation sociale et socioculturelle*, because the facilitators were caught in the middle of daily altercations and violence from young adults who are constantly in conflict with each other. He explained that the premises and the 'institution' of the *Maison des Jeunes et de la Culture* (*MJC*) has the place it has today thanks to the strong symbolism of its history and the attachment of the inhabitants to this place. This being said, he insisted on the fact that if tomorrow the young people came 'to break the premises and the windows' of the structure, that it did not matter to him because the *MJC* had a 'concrete insurance.' With this, he explains that the team relies too much on the premises, which have a symbolic force, but that if tomorrow these no longer exist, the essence of their animation should be able to exist in the street. This sentence marked me because it reveals a certain indifference to space in animation and more precisely in the professional experiences that I have had so far. It was as if the essence of popular education found a certain strength of meaning in the public space and outside the walls.

I immediately made the link with *Groupement Rennais de Pédagogie et d'Animation Sociale* (*GRPAS*), in which I carried out my *Service civique*. This outside animation, which can only exist with the consistency of the unforeseen, of the random, of the spontaneity of the public, is the one that has the most meaning in my eyes. The apprehension of the space by the pedagogues of the *GRPAS* is, in my opinion, part of the ideology of integration. Indeed, it: 'is articulated around mainly 'non local' spatial references: agglomeration, continent, world. But above all, it conveys a new definition of space, a definition that leads those who share it to focus less on the particular characteristics of places than on the forces that make them uniform' ... as stated in Anne Gilbert's text (1985, 222) (Cherruault, 2022).

11.4 Conclusion: To Move Beyond a Cosmetic Reform

The implementation of the UBT reform is still in its early stages. *LabSET* is due to evaluate its implementation. It is likely that this evaluation will be limited to checking that the branches of the new diploma comply with the main headings of its model of the competency-based approach to training, without going into detail about their content, which will be revised again in 2026. The haste with which this reform was conceived did not allow for any real consultation between teachers on the cross-disciplinary nature of some of their teaching content, whether it be the notion of territory or other elements of the *Carrières sociales* syllabus. Today, the reform's move from a 2-year to a 3-year university degree, with constant human resources, does not facilitate this emergence. Furthermore, the 15% reduction in the number of teaching hours per student per year for the first 2 years of the UBT has not reduced their overall workload, as the number of assessments of their achievements has followed the change from four teaching units to five skills. This does not make it any easier for students to assess the cross-disciplinary learning that this reform was supposed to promote.

This consultation would, therefore, be necessary to provide a valid basis for the curriculum standard in which the teaching of the geographical dimension of the concept of territory can find its place. It should also be able to lead to a concept of training in the *Carrières sociales* branches which, while taking into account the notion of skill, does not limit itself to explaining its meaning in a hurry, and without providing for the possibility of its regular revision, leading to random categorizations (see Sect. 11.2), but entrusts its teachers with the responsibility of its progressive development.

References

Bergougnan, R., & Fondeville, F. (2021). Social work education and training in France: A long history of being energised by an academic discipline and international social work. In M. Laging & N. Žganec (Eds.), *Social work education in Europe. Traditions and transformations* (pp. 65–88). Springer Cham. https://doi.org/10.1007/978-3-030-69701-3_4

Boucher, M., Sentis, F., Berthet, T., Dansac, C., Alix, J.-S., Gallasse, J.-P., Gilbert, Y., Baldelli, B., Marchandet, E., Roulhac, C., & Fofana, P. (2024). Établissements de Formation en Travail Social et Universités à l'épreuve de l'universitarisation (de quoi s'agit-il ? Quels sont les problèmes et contradictions ?). *Sciences et actions sociales, 21*. http://journals.openedition.org/sas/4151

Cadoret, L. (2022). *S'initier aux démarches d'enquête ou de diagnostic de territoire*. LAS Introduction to territorial survey and/or diagnosis, Université de Rennes.

Camus, J., & Lebon, F. (2015). *Regards sociologiques sur l'animation*. La Documentation française.

Chauvière, M., & Gaillard, R. (2020). Formations au travail social et universités. Sociologie d'une asymétrie historique devenue complicité adverse. In J.-S. Alix (Ed.), *Le travail social en quête de légitimité* (pp. 73–111). Presses de l'EHESP.

Cherruault, M. (2022). *S'initier aux démarches d'enquête ou de diagnostic de territoire*. LAS Introduction to territorial survey and/or diagnosis, Université de Rennes.

Debarbieux, B. (2014). Enracinement - Ancrage - Amarrage: raviver les métaphores. *L'Espace géographique, 43*, 68–80. https://doi.org/10.3917/eg.431.0068

Georges, F., & Poumay, M. (2020). *Rédiger le référentiel de compétences du Bachelor Universitaire de Technologie*. https://apliut.fr/wp-content/uploads/2020/06/Guide_redaction_referentiel_competences_LABSET_ADIUT-1.pdf

Gilbert, A. (1985). Et si les géographes s'intéressaient aux idéologies moins officielles. *Cahiers de géographie du Québec, 29*(77), 217–224. https://doi.org/10.7202/021719ar

Giraut, F. (2008). Conceptualiser le territoire. *Historiens et Géographes, 403*, 57–68.

Greffier, L. (2013). *Animer le territoire. Territorialiser l'animation*. Carrières Sociales Editions.

Grek, S. (2024). *The new production of expert knowledge education, quantification and Utopia*. Palgrave Macmillan.

Groupe de Travail Commission Pédagogique Nationale / Commission Consultative Nationale des Instituts Universitaires de Technologie. (2020). *Écriture des programmes nationaux des mentions de bachelors universitaires de technologie*. https://apliut.fr/wpcontent/uploads/2020/06/NOTE-de-CADRAGE-BUT-VF10_séance_postCCN_18_juin_2020.pdf

Lafortune, J.-M., Augustin, J.-P., Bélanger, P., & Gillet, J.-C. (2010). Vers un système d'animation socioculturelle : défis actuels et synergies internationales. *Animation, territoires et pratiques socioculturelles, 1*, 1–12. https://doi.org/10.55765/atps.i1.80

Lebon, F. (2020). *Entre travail éducatif et citoyenneté : l'animation et l'éducation populaire*. Champ social.

Ministère de l'Enseignement supérieur, de la Recherche et de l'Innovation (2022). *Bachelor Universitaire de Technologie Carrières Sociales Parcours ASSC-AS-CGE3S-ES-VTD*. https://www.enseignementsup-recherche.gouv.fr/sites/default/files/imported_files/documents/bospe-4-26-5-2022-142651 7pdf-18506.pdf

Moine, A., & Sorita, N. (2015). *Travail social et territoire. Concept, méthode, outils*. Presses de l'EHESP *Motion de l'Assemblée des Chefs de Départements Carrières sociales (19 novembre 2019)*. https://www.snesup.fr/article/motion-de-lacd-carrieres-sociales

Piou, A. (2022). *S'initier aux démarches d'enquête ou de diagnostic de territoire*. LAS Introduction to territorial survey and/or diagnosis, Université de Rennes.

Thémines, J. (2016). Propositions pour un programme d'agir spatial : la didactique de la géographie à l'épreuve de changements curriculaires. *Les Sciences de l'éducation - Pour l'Ère nouvelle, 49*, 117–150. https://doi.org/10.3917/lsdle.494.0117

Régis Keerle is Lecturer at Social Careers' department in the University of Rennes, and Researcher at ESO-Rennes, Spaces and Societies' Laboratory (UMR CNRS 6590 ESO) in the University of Rennes 2. From an epistemological perspective of interdisciplinary cumulativity of knowledge, his geography research focuses in particular on sport, trade unionism and more generally on services to populations (including education), with a theoretical focus on spatialization and territorialities, and a methodological focus on mapping.

Chapter 12
A New Curriculum for Teaching and Learning Geography in French Elementary School. Some Reflections and Critical Notes

Philippe Charpentier and Fabio Fatichenti

12.1 Introduction

Since the 1990s, France has experienced several renewals of the geography curriculum in primary schools (1995, 2002, 2007, 2008, and 2015) (Chevalier, 2008; Philippot & Charpentier, 2016). Roumégous (2002) pointed out that the aim of geography in elementary schools is to foster a better understanding of the world without being conditioned by one's own representations, based on simplistic explanations of causality (Roumégous, 2002). However, Philippot (2008) showed, as early as 2008, that geography taught in elementary school was mainly focused on disciplinary markers (presence of maps, location, and naming of places).

Anyway, several scholars specialized in geography teaching and learning (Thémines, 2006; Roumégous & Clerc, 2008; Meunier, 2018; Colin et al., 2019; Gaujal, 2019; Filâtre, 2021; Charpentier & Fatichenti, 2023) have made proposals to give life and meaning to past and current French curricula. For instance, Meunier suggests using children's literature to introduce pupils to geographical issues, while Filâtre proposes that pupils should study their school's immediate environment, and Charpentier and Fatichenti that geography textbooks do not simply ask students factual questions, but also stimulate their critical thinking skills.

Even if we do not expect elementary school pupils to become professional geographers (Le Roux, 1997), we postulate that elementary school geography, like other

P. Charpentier (✉)
Université de Mayotte, Dembeni, France
e-mail: philippe.charpentier@univ-mayotte.fr

F. Fatichenti
Università degli Studi di Perugia, Dipartimento di Lettere, Perugia, Italy
e-mail: fabio.fatichenti@unipg.it

M. Hardouin, C. Leininger-Frézal (eds.), *Curricular Changes in Geography Education*, International Perspectives on Geographical Education,
https://doi.org/10.1007/978-3-031-96551-7_12

school subjects, should be able to encourage pupils to reflect and analyze the world around them, starting from engaging and motivating activities.

To do so, the geography taught in elementary schools has to be organized around a disciplinary matrix (Develay, 1992) and employ its own terminology, questioning, methods, and tools (Giorda, 2014; De Vecchis et al., 2020).

Knowing this, we propose in the following sections a new way of critical thinking about the geography curriculum in French elementary schools. The aim here is not to propose a complete curriculum for all levels of elementary school but to think differently about the aims of future curricula. In other words, in this chapter, we aim to introduce guidelines for the development of a new possible future curriculum that could change the way geography is currently taught.

The next section will describe the underlying theoretical framework. It will define the objectives and expectations of this curriculum. Section 12.3 will seek to answer methodological questions. Section 12.4 will deepen a discussion about the interest of this new curriculum, the role played by the teaching of geography facing the Rs tryptic, and the relevance of geographic questions related to students' daily lives.

12.2 Theoretical Framework

12.2.1 A Curriculum for the Children of Today and the Adults of Tomorrow and for Non-Specialist Elementary School Teachers

In agreement with Meirieu (2021), we believe that, before creating the contents of a curriculum, it is necessary to reflect on its purpose, on the kind of human we want to see emerge from the evolution of the child into an adult. The primary goal we ascribe to school is to enable pupils to become aware, emancipated adults (Charpentier, 2023) able to observe, question, seek information, and argue using appropriate disciplinary terminology. But this goal only makes sense if pupils are aware or become aware that the world in which they live can be only inspired by solidarity (Jacquard, 1999), democracy (Meirieu, 2021), and the common good (Ogien, 2007).

Bearing in mind that French elementary school teachers do not receive in-depth training in all the subjects they teach, as might be the case of French secondary school teachers (Baillat, 2001; Philippot, 2008), the idea that we suggest in this chapter is to develop a curriculum that does not cause difficulties to them, particularly in terms of the elaboration of the teaching-learning activities they create for their pupils. We have taken up the distinction made by the French ergonomic school between *task* and *activity* (Vinatier & Pastré, 2007). The task is what the curriculum proposes to do in class and the activity that the teacher actually makes in class with the pupils. But if we want elementary school teachers to teach geography effectively

(Philippot & Baillat, 2009), if we want them to propose motivating activities to their pupils, and if we want them to learn the curriculum topics as they were conceived by their authors, it seems necessary to develop a link between the curriculum to be taught and the activities realized (Leplat, 1992). As stated by Philippot and Baillat (2009), many French elementary school teachers do not respect the time assigned to teaching geography or the proposed curriculum (this is also the case for other subjects such as sports and the arts, for example). In fact, as it has been observed by the authors of this chapter, they spend more time teaching mathematics and French. Charpentier (2018), for his part, showed that some of them show difficulties in understanding the curriculum that is supposed to be taught during the geography lessons.

Hence, if we want to change the currently available curriculum for the teaching of geography in elementary school, we need to take these conditions into account. Thus, we postulate that these proposals must be considered reasonable if they are feasible (Charpentier et al., 2023). On the basis of Vygotski's studies (1934, 1978), we recall the constructivist concept that mastering knowledge means first and foremost incorporating it into oneself experience, in a movement coming from things or events that are external to human beings (Pastré, 2007). In other words, mastering knowledge does not just mean repeating it by heart, but understanding it, reflecting on it, and incorporating it into personal experiential paths. So, we think it is important that the teaching curriculum for both teachers and pupils must be as precise, flexible, and explicit as possible (Leplat & Hoc, 1992).

From the book "*School culture in geography, The world in the classroom* (fr. *La culture scolaire en géographie, Le monde dans la classe)*" by Pascal Clerc in 2002, which deals with textbook proposals in relation to the secondary school programs of the time, we postulate that *it is not the world that must enter the classroom but the classroom that must enter the world.* In other words, it is pupils' questioning about the surrounding world that should take up the main place in the curriculum. We might say then that everything that surrounds pupils could and should become a source of geographic inquiry. As Bartlett (quoted by Vergnaud, 2002) assumes, each discipline enables its users to question the world around them in a specific and contextual way and to give it a definite meaning by producing a discourse on it.

12.2.2 The Aims of a Possible Renewed Curriculum

12.2.2.1 Understanding That Spaces and Territories Are Interconnected

Now more than ever space and territory are no longer independent of each other but interconnected (Perez & Sokolov, 2020) in a globalized world (Gélinas, 2007). In order to observe and analyze the surrounding reality, it would be advisable not to adopt a too narrow or too broad perspective. Better, an integrated view of these two dimensions of reality would be needed to interact effectively with the external world (De Rosnay, 2014).

12.2.2.2 Developing Strategies and Exploiting Specific Language for Learning Geography in Elementary School

We postulate that the teaching of geography should enable students to acquire a specific terminology, strategies, and tools to be used by pupils but we must take into account their individual ways of learning and exploiting what is offered to them in class. The aim of this curriculum is to teach pupils how to learn actively and not to repeat by heart knowledge passed on unilaterally by teachers or media.

According to us, geography should enable pupils to think autonomously, develop an informed and critical vision of the world around them, and act consciously in the present and future. In other words, it is the teacher's support in the development of an oriented vision of a pupil, considered as a future emancipated actor (Charpentier, 2023), that we are ultimately aiming for through our proposal.

This latter takes into account not only the form of schooling (Vincent et al., 1994) and the age of the pupils but also the context. In addition to the knowledge to be taught, the school form includes a space and time dedicated to the subject. We are conscious that the time devoted to teaching geography in schools is increasingly restricted; according to the latest curricula mentioned above, it extends to a maximum of one-hour sessions per week.

The age of the pupils also needs to be taken into account when designing curricula that have to be progressive and accessible to them while at the same time being demanding in terms of learning.

12.2.2.3 Learning Less but Better

We believe that what is important is not so much the quantity of the acquired knowledge, but rather the quality of the intellectual and didactic approach proposed. Indeed, if we ask too much to teachers and pupils, we run the risk that what is planned will not be carried out and the pedagogical objectives will not be achieved. Following the example of Francesch (2011) and Zavalloni (2022), we advocate a pedagogy of slowness that allows pupils to take their time to learn.

In this respect, we believe that it would be more than sufficient for pupils to relate to a single geographic macro issue for each school period, in order to become able to engage in a meaningful intellectual and learning process. This would give teachers the time they need to develop all aspects of this topic in terms of documentary research, knowledge, methods, tools, and argumentative production by the pupils.

12.2.3 A Geography Curriculum Built Around Three Distinct Yet Interconnected Functions

We do think that the geography curriculum has to articulate together questions, languages, methods, and geographic approaches together to build a meaningful and involving system. Following Yatchinovsky (2018), we define a system as a set of elements or individuals, of which the one under study is a part, and of which all the parts that make it up are interdependent. This systemic approach is of interest to us because it articulates closely:

- The questions and documentary research that pupils will have to carry out according to their age and literacy level.
- A common culture that includes shared reference points.
- The opportunity for students to project themselves into the present and future by developing a reasoned geographic discourse to answer the question posed.

Each part is linked to the others and can never be separated from them. In fact, these three elements constitute a system; they can only make sense in relation to each other, as they complement each other. By asking questions and finding ways of answering, pupils acquire languages (Bentolila, 2017), tools, and methods (La Borderie, 1991) specific to geography, and they will be consequently able to give meaning to what they do in class and what they experience inside and outside the school, to speak the same language in order to observe, to question, to understand, to understand each other, and to propose actions for today and tomorrow on the spaces and territories that are close to their life experience or far away.

Under these conditions, the aim of elementary school geography will be to give pupils the opportunity to step outside their own world (Rabachou, 2016).

12.2.4 A Curriculum That Looks at Each Student as a Unique Subject and an Actor

In this curriculum, the student is seen as not only a subject but also an actor, but one who is inscribed in the society in which he or she lives (Durkheim, 2022), a society that nowadays is globalized (Gélinas, 2007) and interconnected (Perez & Sokolov, 2020). But each pupil, even if he or she is an individual (Rabachou, 2017), is, nonetheless, a member of a society that is connected to people other than himself or herself (Lévinas in Hocquard, 1996), other societies (Barthelemy, 2023), a de facto world of solidarity (Jacquard, 1999; Jankélévitch, 1984).

12.3 How to Implement This Possible New Curriculum?

12.3.1 Enabling Pupils to Investigate, Question, and Document

We believe that geography is not a dead science, is not just of matter of "general culture" or a science that is useful only at school with the aim of knowing a list of names of capitals, mountains, rivers, and so on, or to pass a test (Philippot & Baillat, 2009) and move on to the next grade (Freire, 2018; Meirieu, 2021).

Geography is, instead, a living science that questions the world, aware that each human being is, by nature, a geographical being (Joublot-Ferré, 2018). And human beings are obliged, somehow, to be interested in what surrounds them, in the way in which their surroundings are spatially organized, whether near or far. Nowadays, we are surrounded by and living with objects from all over the world. Learners are entering an already organized society, in which they are rarely asked to consider what is closest to them, the spatial organization of their lives, of their society, and of the world, even though it is interconnected and globalized. According to us, enabling pupils to observe the world around them, to question it, and to ask questions about it means to enable them to understand it better.

12.3.2 A Curriculum Organized Around Pupils' Emerging Questions About the World Around Them

Questions lead to attempts to find answers and a possible development of language, methods, and tools to solve them. This questioning can be carried out by the pupils themselves, or supported by the teacher, bearing in mind that, as Rousseau put it back in 1762 in L'*Emile ou De l'éducation* (2009), the role of the teacher is to enable pupils to ask themselves questions, to discover autonomously what we want them to discover, and to accomplish the necessary steps to understand it.

To answer their questions, the geography sessions should then enable pupils to carry out documentary research using the media currently available (whether they are helped by their teachers, depending on their class level) and reading documents (whatever their nature and function). This first part of this three-stage system is a prerequisite for enabling pupils to acquire a specific lexicon, common reference points, and elements for arguing geographical answers to the question they have developed.

12.3.3 A Curriculum to Build a Shared Culture Based on a Common Language and Geographical References

Geography helps to organize the world (terrestrial space). Historically, this was one of its first functions (Bavoux, 2016): naming, localizing, and finding one's bearings.

In addition to toponyms to remember and geographic landmarks to know, these landmarks also include terms, a geographic lexicon (landmarks, terms for locating places, people, and objects in relation to one another) that help people to share a common culture, a culture that, at the very least, enables us to understand one another, to understand what we are talking about, and to know which place we are talking about and which place we are questioning.

12.3.4 A Curriculum Aimed to Develop a Reasoned Geographic Discourse

The aim of this curriculum is to enable pupils, on the basis of the questions they ask themselves as a result of their observations of the world around them (whether these questions concern phenomena or objects that are near or far), to think, to represent, to propose a discourse, to project themselves, to criticize, and to develop an informed view of it. Since pupils have observed what surrounds them, that they question it or are led to question it by their teachers, that they know what it is, because they can name it, locate it, and define it, and that they have done research on it (with or without the help of their teachers), then they can develop a reasoned perspective and propose a geographic discourse (which can be oral, written, drawn, diagrammed, mapped, etc.) on the issues it faces. The aim is to make pupils fully active in the classroom and in the world around them, according to their age and class level, and give them the opportunity to express themselves in a critical way exposing their point of view.

12.4 Discussion

12.4.1 Reasoned and Reasonable Proposals Based on the Aims of the Renewed Curriculum

Given the time constraints inherent to the teaching of geography in the classroom, and the necessity for teachers to comprehend the curriculum in order to offer pupils activities that meet its objectives, so that pupils can learn with confidence, the following suggestions are put forward.

Let us begin with the principle laid down by Dewey (1931) that any teaching curriculum, whatever the discipline, must answer questions posed by the society in which the pupils live. It seems to us that geographical curriculum should not be built around topics to be studied, but rather around pupils' needs and personal experiences. In 1931, Claparède advanced the argument that education should prepare children for life, but that it should also prepare them for life in the present, while embodying life itself.

But how can this education be brought to life, in touch with the pupils' present life, if not by enabling them to question what surrounds them, what is closest to them, or by starting from the questions of the society in which they are embedded? This would offer the possibility to enable them, little by little, to develop a geographical approach.

As an example, and based on a concrete situation taken from the pupils' daily lives, the fact of questioning school equipment, its origin, its method of manufacture, its cost, and so on would enable them to enter into a de facto geographic questioning, to take an interest in what surrounds them, and to give meaning to the discipline by being in touch with their daily lives and objects.

If the pencil they are using is made of wood and contains a graphite lead, then the question is: What is wood, what type of wood can be used to make pencils, and where can they be produced? Furthermore, what production methods can be used? Where are the pencils actually produced? And what about the graphite it contains and which is used for writing? What is it made of? Where is it found? And what about the packaging that surrounds the purchased pencil, the organization of the distribution chain between producer(s) and buyer(s) (the communication networks required for this distribution, the storage facilities, the intermediaries, the management of waste linked to the production of the product and its end-of-life, etc.)? The same type of questioning could just as easily be based on the food eaten by pupils at breakfast (Grataloup, 2017) or during the day, the journeys and means of transport they may use to get from one place to another, questions concerning the layout of their classroom, their school, and so on. In other words, a set of questions that enable the pupils to enter into a simple geographical reasoning process, based on questions about everyday life, based on what we do not question, or no longer question because it seems implicit. But it seems to us that nothing is self-evident. Living and, for some, surviving today is no more self-evident than it was in the past and will be in the future. Our survival can only take place because what makes it possible is organized and thought out. And the society of today and tomorrow will only be what we want it to be.

Be that as it may, we (the students) can have a documented, argued, critical discourse on what surrounds us (them), so that, if necessary, we (they) can think about the current or future situation differently.

In the curriculum we are proposing, it is no longer compulsory for pupils to study certain themes, as is currently the case in the current curriculum. Instead, the emphasis is on a geographical approach based on questions that pupils are asked about their immediate environment.

12.4.2 The Part Played by the Teaching of Geography in the Face of the Rs Tryptic

The question of teaching geography in elementary school may seem secondary to the three Rs tryptic; reading, writing, and arithmetic, so often put forward by politicians (Bayrou, 1992; Darcos, 2000), editorialists, and intellectuals (Julliard, 2015). And yet, what use are these three activities without an application to a specific disciplinary field? Reading, writing, and counting only have meaning and interest if they are applied to a question, whatever it may be. But beyond their interest, they must serve a purpose, other than a utilitarian one (learning the curriculum as it were designed by others, learning it to pass to the next grade, to obtain diplomas …). The curriculum must be congruent with the type of adult the institution aims to produce. The question is whether we want to produce adults who are used to repeating what they are told or what they have been taught, or do we want adults who are capable of asking questions, reflecting, documenting, arguing, proposing, and having an informed and even critical view of the society in which they live or in which they will be proposed to live?

We prefer the second option, that of future adults having a critical view and operating a sharp analysis of the surrounding world because we believe that the ultimate purpose of school is to enable children to become emancipated adults.

12.4.3 Questions to Be Related to Students' Daily Lives Inside or Outside the Classroom

We believe that curricula should be organized around specific questions (and these questions can be formulated by pupils or proposed by teachers): these interrogate pupils about their surroundings and ask them to notice what we no longer see because we are no longer used to observing.

On the other hand, if we agree that it is the classroom that must enter the world, and that pupils must be encouraged to step outside their own world in a globalized society, how can we effectively develop the relationship between geography and civic education (Fatichenti, 2023) if not with this kind of approach? And what is the point of an approach that does not force us to ask questions about the challenge of sustainability (e.g., Agenda 2030) or anything else we are used to understanding in the crucial issue of the Anthropocene (Charpentier & Fatichenti, 2023)?

Nor should we overlook the spin-offs that these proposals can bring by rediscovering methods of teaching geography, such as field trips or all the outdoor activities that pedagogues—think of Jean-Jacques Rousseau (1792/ 2009), Célestin Freinet (1974), or Mario Lodi (2022)—emphasize as indispensable.

The world around us is a source of questions, geographic questions that call for the use of a language, geographic landmarks to build understanding, methods, geographic tools, documents to mobilize, and a reasoned proposal to report on what we have understood. But these different aspects only make sense if they form a system, that is, if they are interdependent on each other, if they respond to and complement each other to produce an argumentative geographical discourse on the world around the pupils; whether this world is near or far. But these reasoned geographical responses to these questions only make sense if we envisage pupils grappling with a world of de facto solidarity, a world in which the decisions of some can have consequences for the lives of others, a democratic world in which the lives of some are not built to the detriment of others, but are developed with the common good in mind.

12.5 Conclusion

This reasoned proposal for a renewed curriculum for French primary schools is based on the fact that, in many classes, geography is taught little or nothing, and that teachers offer often low cognitive level activities to their pupils: naming places, locations, descriptions, and so on. The current geography curriculum requires teachers to address pre-set subjects.

Nowadays, whether we like it or not, we are surrounded by everyday objects that we no longer see because they are constantly under our gaze and that, moreover, we do not question with a geographical eye (what are they made of? by whom? how? where?…). These are the questions pupils should ask themselves and then answer, using geographic language, tools, methods, and reasoning.

It is suggested that the curriculum be renewed by proposing a method to be followed, rather than mandatory topics to be studied. Indeed, it is believed that it is from the questions that arise the usefulness of using the geographic approach. This approach empowers pupils to adopt and put into practice a particular way of thinking and problem-solving that addresses the world around them.

References

Baillat, G. (2001). *Enquête sur la polyvalence des enseignants du premier degré*. Rapport de recherche. IUFM de Reims Champagne-Ardenne.

Barthelemy, M. (2023). *Le monde des réseaux*. Odile Jacob.

Bavoux, J.-J. (2016). *La géographie – Objets, méthodes, débats*. Armand Colin.

Bayrou, F. (1992). *1990–2000, La Décennie des mal-appris*. Flammarion.

Bentolila, A. (2017). *L'école contre la barbarie*. First Editions.

Charpentier, P. (2018). Le cheminement intellectuel et pratique d'un maître non spécialiste confronté à la préparation de séances de géographie au cycle 3 de l'école primaire. *L'Information géographique, 3*(82), 132–143.

Charpentier, P. (2023). Des savoirs émancipateurs dans une école émancipatrice ? Quelques réflexions philosophiques et pédagogiques. In J.-C. Buttier & A. Panagiotounakos (Eds.), *Des savoirs pour agir sur le monde. Quels apprentissages des élèves face aux enjeux contemporains ?* (pp. 203–218). PUG.

Charpentier, P., & Fatichenti, F. (2023). Pour une éducation géographique à l'Anthropocène à l'école primaire en France et en Italie: situation actuelle, critique et perspectives. *L'Information géographique, 87*(1), 93–113.
Charpentier, P., Frézal-Leininger, C., Genevois, S., Leroux, X., Thémines, J.-F., & Vergnolle Mainar, C. (2023). Contexte et conditions de pratiques de savoir émancipatrices à l'Ecole, à partir du cas de la géographie scolaire française. In J.-C. Buttier & A. Panagiotounakos (Eds.), *Des savoirs pour agir sur le monde. Quels apprentissages des élèves face aux enjeux contemporains ?* (pp. 25–50). PUG.
Chevalier, J. P. (2008). Enseigner la France en géographie aux jeunes écoliers (1788-2008). *L'Information géographique, 72*(3), 20–33.
Claparède, E. (1931). *L'éducation fonctionnelle*. Delachaux et Niestlé.
Clerc, P. (2002). *La culture scolaire en géographie. Le monde dans la classe*. PUR.
Colin, P., Heitz, C., Gaujal, S., Giry, F., Leininger-Frézal, C., & Leroux, L. (2019). Raisonner, raisonnements en géographie scolaire. *Géocarrefour, 93*(4). https://doi.org/10.4000/geocarrefour.12524
Darcos, X. (2000). *L'Art d'apprendre à ignorer*. Plon.
De Rosnay, J. (2014). *Le Macroscope: Vers une vision globale*. Points.
De Vecchis, G., Pasquinelli, D., & Pesaresi, C. (2020). *Didattica della Geografia*. UTET.
Develay, M. (1992). *De l'apprentissage à l'enseignement*. ESF.
Dewey, J. (1931). *L'école et l'enfant*. Delachaux et Niestlé.
Durkheim, E. (2022). *Education et Sociologie*. Puf.
Fatichenti, F. (2023). Una Geografia per la sostenibilità. Il quadro epistemologico disciplinare, le Indicazioni nazionali, la prassi didattica nella Scuola primaria. *Annali online della Didattica e della Formazione Docente, 15*(25), 126–140. https://doi.org/10.15160/2038-1034/2634
Filâtre, E. (2021). *Développer la conscience géographique des élèves en enseignant à partir de l'espace proche*. Thèse de doctorat en géographie. Toulouse 2-Jean Jaurès.
Francesch, J.-D. (2011). *Éloge de l'éducation lente*. Chroniques sociales.
Freinet, C. (1974). *Pour l'école du peuple*. Maspéro.
Freire, P. (2018). *Pedagogy of the oppressed*. Bloomsbury Academic.
Gaujal, S. (2019). La cartographie sensible et participative comme levier d'apprentissage de la géographie. *VertigO - la revue électronique en sciences de l'environnement, 19*, 1. https://doi.org/10.4000/vertigo.24604
Gélinas, J.-B. (2007). *La Globalisation du monde*. EcoSociété.
Giorda, C. (2014). *Il mio spazio nel mondo. Geografia per la scuola dell'infanzia e primaria*. Carocci.
Grataloup, C. (2017). *Le monde dans nos tasses. Trois siècles de petit déjeuner*. Armand Colin.
Hocquard, A. (1996). *Eduquer, à quoi bon? Entretien avec Emmanuel Levinas*. Puf.
Jacquard, A. (1999). *Petite philosophie à l'usage des non-philosophes*. Le Livre de Poche.
Jankélévitch, V. (1984). *Le Paradoxe de la morale*. Seuil.
Joublot-Ferré, S. (2018). Introduction. Faut-il encore enseigner la géographie à l'école? *L'Information géographique, 82*(3), 10–14.
Julliard, J. (2015). *L'école est finie*. Flammarion.
La Borderie, R. (1991). *Le métier d'élève*. Hachette.
Le Roux, A. (1997). *Didactique de la géographie*. PUC.
Leplat, J. (1992). L'analyse psychologique du travail. In J. Leplat (Ed.), *L'analyse du travail en psychologie ergonomique* (pp. 23–39). Octares.
Leplat, J., & Hoc, J.-M. (1992). Tâche et activité dans l'analyse psychologique des situations. In J. Leplat (Ed.), *L'analyse du travail en psychologie ergonomique* (pp. 47–60). Octares.
Lodi, M. (2022). *Cominciare dal bambino*. Rizzoli.
Meirieu, P. (2021). *Dictionnaire inattendu de pédagogie*. ESF.
Meunier, C. (2018). Quand les albums parlent d'espace. Enseigner la géographie avec le livre pour enfants. *L'Information géographique, 82*(3), 50–71.
Ogien, R. (2007). *L'éthique aujourd'hui: Maximalistes et minimalistes*. Folio.

Pastré, P. (2007). Quelques réflexions sur l'organisation de l'activité enseignante. *Recherche et formation, 56*, 81–93. https://doi.org/10.4000/rechercheformation.907

Perez, C., & Sokolov, K. (2020). *Un monde en réseau: Initiation par la pratique à la théorie des graphes*. Independently Published.

Philippot, T. (2008). *La professionnalité des enseignants de l'école primaire. Les savoirs et les pratiques*. Thèse de Doctorat. Sciences de l'Éducation, IUFM de Reims.

Philippot, T., & Baillat, G. (2009). Les enseignants du primaire face aux matières scolaires: Réflexions sur la professionnalité enseignante. *Recherche & formation, 60*, 63–74.

Philippot, T., & Charpentier, P. (2016). Les mutations de la géographie scolaire à l'école primaire en France. Entre prescriptions et pratiques d'enseignement. *Spirale—Revue de recherches en éducation, 58*(2), 121–132.

Rabachou, J. (2016). *Qu'est-ce qu'un monde ?* Vrin.

Rabachou, J. (2017). *L'individu reconstitué*. Vrin.

Roumégous, M. (2002). *Didactique de la géographie. Enjeux, résistances, innovations 1968–1998*. PUR.

Roumégous, M., & Clerc, P. (2008). Le bon sens tout près de chez soi: les projets de programme de géographie pour l'enseignement primaire, *Cybergeo: European Journal of Geography*, Débats, Les nouveaux programmes dans le primaire. http://cybergeo.revues.org/20383.

Rousseau, J. J. (2009). *Emile ou De l'éducation*. Flammarion.

Thémines, J.-F. (2006). *Enseigner la géographie: un métier qui s'apprend*. Hachette.

Vergnaud, G. (2002). Piaget visité par la didactique. *Intellectica - La revue de l'Association pour la Recherche sur les sciences de la Cognition, 33*, 107–123.

Vinatier, I., & Pastré, P. (2007). Organisateurs de la pratique et/ou de l'activité enseignante. *Recherche et Formation, 56*, 95–108.

Vincent, G., Lahire, B., & Thin, D. (1994). Sur l'histoire et la théorie de la forme scolaire. In G. Vincent (Ed.), *L'éducation prisonnière de la forme scolaire ?* PUL.

Vygotsky, L. S. (1934, 1978). *Mind in society. The development of higher psychological processes*. Harvard University Press.

Yatchinovsky, A. (2018). *L'approche systémique: Pour gérer l'incertitude et la complexité*. ESF.

Zavalloni, G. (2022). *La pedagogia della lumaca. Per una scuola lenta e nonviolenta*. EMI.

Philippe Charpentier is Associate Professor in Educational Sciences at Mayotte University (France) and a permanent researcher of ICARE laboratory at the University of La Réunion. His research focuses on the activities of pupils and teachers in primary school, the teaching and learning of geography, environmental and sustainable development issues in primary schools, and more generally on the meaning of school for the society at large, teachers and pupils.

Fabio Fatichenti is Associate Professor in Geography at the University of Perugia (Italy). His research activity has resulted in approximately 140 publications (monographs and articles published in journals and proceedings of national and international conferences) focusing particularly on themes and issues related to landscape, environment, and territory, the study and enhancement of cultural heritage, as well as the teaching of Geography and the analysis of geo-historical sources.

Chapter 13
Pupils as Spatial Citizens: Design Principles for Education for Sustainable Development Projects to Promote Pupils' Participation

Lisa Wieczorek, Nina Scholten, and Natalie Bienert

13.1 Setting the Scene: Striving for Common Change in an Increasingly Complex World

Climate change, its impacts, and its consequences pose and will continue to pose significant challenges, affecting urban design, living conditions, and the well-being of citizens in a variety of ways. Addressing these issues requires a socio-ecological transformation, which requires collective efforts to identify sustainable ways of living (Brand & Wissen, 2017). This also includes strengthening a culture of participation.

Considering the various, broadly defined definitions, participation involves active engagement in civil and political activities to influence and change societal conditions (e.g., Bruselius-Jensen et al., 2021; Israel et al., 2021). In this context, it seems important to actively include the concerns and perspectives of young adults. Despite frequent calls for youth participation in urban planning, it is rarely implemented in practice (Rieker et al., 2016), which is problematic, as young adults are often equally, if not more, affected by the aforementioned trends. Excluding their perspectives denies urban planning projects their potentially important contributions to the sustainable development of urban life (Ataol et al., 2019).

L. Wieczorek (✉) · N. Bienert
Institute for Geography Education, University of Münster, Münster, Germany
e-mail: lisa.wieczorek@uni-muenster.de; natalie.bienert@uni-muenster.de

N. Scholten
Institute for Geography, University of Hildesheim, Hildesheim, Germany
e-mail: nina.scholten@uni-hildesheim.de

M. Hardouin, C. Leininger-Frézal (eds.), *Curricular Changes in Geography Education*, International Perspectives on Geographical Education,
https://doi.org/10.1007/978-3-031-96551-7_13

Young adults' exclusion is often rooted in a widespread perceived lack of participatory competencies among adults (Rieker et al., 2016). Participatory competencies are defined as the ability to gain knowledge about various forms and opportunities for participation and to utilize them effectively; this encompasses the capacity to articulate and express personal interests, share them, and contribute to public discourses (Staatsinstitut für Schulqualität und Bildungsforschung München, n.d.). Strengthening these competencies seems essential to fully benefiting from young adults' perspectives in planning processes. To facilitate their development, it is essential to implement schools in these processes; they are pivotal institutions for education and socialization and aim to prepare young adults to become responsible citizens, which encompasses the fostering of participatory competencies. Especially geography education, as a discipline focused on human–environment interactions, with its diverse methodological approaches, strong relevance to everyday life, and opportunities to promote and initiate direct action-oriented learning, appears essential for fulfilling this educational mandate (Hinde, 2018). This is also reflected in the core frameworks of the discipline, such as transformative learning (UNESCO, n.d.) and education for sustainable development (ESD). Nevertheless, research suggests that the prevailing teaching approaches in ESD, transformative learning, and also political-geographical education are not entirely consistent with this educational mandate (Gryl & Budke, 2016; Jo, 2018). In conclusion, it is imperative to develop, establish, and cultivate learning environments, especially in geography education to foster young adults' participation. This study addresses the question of how a participatory project can be designed to achieve the aforementioned objectives of promoting participatory competencies while engaging young adults more effectively in real-life planning issues.

To address this question, Sect. 13.2 outlines the main theoretical foundations and current trends in participation practices in urban planning, as well as approaches taken to promote participatory competencies in geography education. Section 13.3 presents ten design principles, which are derived from a qualitative literature review for constructing a suitable learning environment. Finally, Sect. 13.4 details the implementation of these principles in a real-life project context.

13.2 Participation as a Foundation for Sustainable Change and Future Development

This section introduces a dual perspective on participation, termed learning and living participation (Fig. 13.1), which are interrelated and mutually dependent. Consequently, a participatory framework should integrate both perspectives, leveraging their inherent possibilities and guiding principles. The key aspects of each approach are briefly outlined below before leading to the formulation of the design principles.

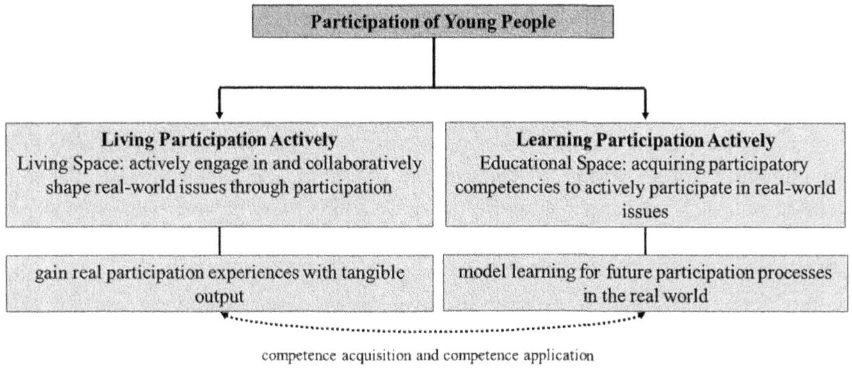

Fig. 13.1 Dual understanding of participation: living and learning participation

13.2.1 Living Participation: Participation in Practice—Forms and Formats

The participation of young adults is becoming increasingly prevalent in political and societal contexts. However, there is no consensus or theoretical framework that defines or guides its implementation (Bruselius-Jensen et al., 2021). Consequently, various participation formats exist. To develop design principles for the creation of a successful participation environment that embodies the dual conception of participation described above, the following formats from urban planning practice can be considered:

Informal participation structures, such as planning cells, offer greater flexibility than planning conferences and other formal structures, which are often quite rigid (e.g., Ohl, 2009; Rieker et al., 2016). In terms of the design and degree of participation, Hart's ladder of participation (1992) provides a key framework for defining the extent of participation among children and young adults. The framework distinguishes eight levels, ranging from "manipulation" to "child-initiated shared decisions with adults" (Hart, 1992, p. 8). Hart emphasizes the importance of avoiding lower levels of pseudo-participation and striving for higher levels, such as the sixth level, at which adults provide a framework that allows children and young adults to contribute freely and shape the project's direction in accordance with their own ideas and needs. However, varying levels of participation are often necessary, incorporating phases of information sharing alongside phases of flexibility and freedom. Therefore, it is essential that participation structures remain flexible to adapt to the individuals involved (Hart, 1992).

Participation formats also vary in terms of the initiator and the determination of the topics and questions addressed. It is preferable for participation initiatives not only to be imposed by politicians (a top-down design; Ohl, 2009) but also to emerge from the people themselves (a bottom-up design; Rieker et al., 2016). Given the limited experiences of young adults, it seems prudent to launch participation initiatives from an intermediary position (e.g., local non-governmental organization (NGO) and youth advocacy office) (ibid.). In terms of formal and methodological design, project-based structures are more effective than representational structures for engaging young adults, as they are time-bound and focus on specific well-defined issues. The former also permits young adults to determine content priorities that align with their interests and needs (Bruselius-Jensen & Nielsen, 2021; Ohl, 2009; Rieker et al., 2016). Finally, digital participation, including the contribution of geodata or participation in online surveys, has become an important part of urban planning. This approach enables broader engagement and the fostering of innovative outcomes (e.g., Allaste & Cairns, 2021; Rieker et al., 2016).

Participation structures to involve young adults should be as follows:
- Project oriented, enabling methodological diversity and a thematic focus.
- Informal to reach a broader range of young adults and allow flexibility in methods and content.
- Focused on the sixth level of the participation ladder, providing a structured project framework while allowing young adults to further develop the project based on their own ideas, skills, and interests.
- Inclusive of digital participation opportunities.

13.2.2 Learning Participation: Geography Teaching in the Context of Education for Sustainable Development and Spatial Citizenship Education

In order to develop an effective learning environment that encourages young adults' participation, it seems beneficial to gain familiarity with the fundamental concepts and frameworks that underpin geographical education. These include approaches such as transformative learning, ESD, and political-geographical education, as these objectives seek to equip pupils with the capacity to address pertinent future questions and engage in decision-making processes that integrate the dimensions of sustainability (Gryl & Budke, 2016; Meyer & Haubner, 2022). Although these questions are quite complex, they can be effectively addressed through appropriate geographical methods. A set of principles (e.g., action orientation) and methodologies (e.g., inquiry-based learning) for creating such learning opportunities, frequently discussed in the literature (Blum et al., 2021; Gryl & Budke, 2016; Meyer & Haubner, 2022; Rieckmann, 2021; Sukuma Arts e.V., n.d.), is provided in Fig. 13.2. These principles should be considered when developing a learning environment that cultivates participatory competencies.

Furthermore, geography education must evolve to accommodate the pervasive presence of digital (geo)media and capitalize on the participation opportunities that these technologies present and decision-making processes, particularly regarding

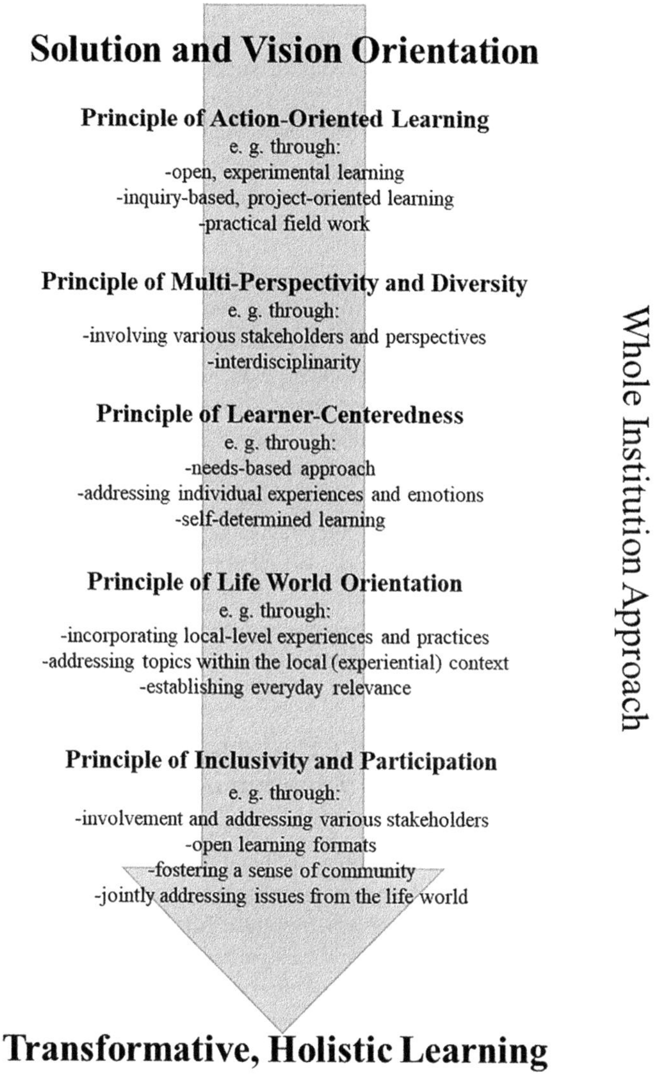

Fig. 13.2 Compilation of principles to teach ESD. Based on the Works of Blum et al. (2021), Gryl and Budke (2016), Meyer and Haubner (2022), Rieckmann (2021), Sukuma Arts e.V. (n.d.)

the responsible utilization of digital and spatial geodata. The so-called spatial citizenship education (SCE) approach emphasizes the acquisition of the knowledge, skills, and critical reflection abilities needed to understand the influences of geospatial data in the digital world and to recognize its implications in daily life. These elements are fundamental to effective participation practices. Beyond merely providing access to geospatial data, the objective of SCE is also to equip pupils with the ability to utilize digital geospatial technologies for active participation and societal transformation through the creation of their own spatial narratives (Gryl & Jekel, 2012; Schulze et al., 2015; Shin & Bednarz, 2018). To cultivate these so-called “spatial citizens”, initial learning approaches, such as service learning (Jo, 2018) and project-based learning (Baker et al., 2018), are discussed in the literature in the context of geography education and SCE. Nevertheless, these approaches remain relatively uncommon in practice and should be more prominently implemented, given their significant potential to enhance participatory competencies within the framework of SCE.

Participation structures designed to promote participatory competencies in terms of the SCE and ESD should perform the following actions:

- Adopting an action-oriented approach, incorporating practical activities in urban spaces with a well-defined structure for content and methodology, and following the principles of inquiry-based learning.
- Embracing the concept of service learning, which integrates various stakeholders and their expertise into the learning process.
- Aligning closely with pupils’ everyday experiences and acknowledging the high thematic complexity and abstract nature of planning procedures.
- Incorporating digital geomedia and recognizing their impactful role in modern participation and decision-making processes.

13.3 Ten Design Principles for Designing School Projects to Promote Pupils’ Participation in Planning Issues

To address the outlined need of integrating living and learning participation in a new participatory framework, ten design principles (see Fig. 13.3) were developed through an extensive qualitative content literature review in which the key points described in Sects. 13.2.1 and 13.2.2 were synthesized. These principles will be employed in a specific project (Sect. 13.4) and evaluated using process-analytical qualitative methods (e.g., interviews) in accordance with the CIPP evaluation model which contains a Content-, Input-, Process-, and Product evaluation (Stufflebeam & Zhang, 2017), with the aim of optimizing the developed framework for further investigation into geography education. The derived principles are intentionally comprehensive, encompassing both general structural considerations and detailed design options. Some principles are closely interrelated, and a complete separation is neither feasible nor desirable, as the objective is to develop a comprehensive and coherent approach. In addition to the principles themselves, suggestions for practical implementation by teachers are provided alongside each principle.

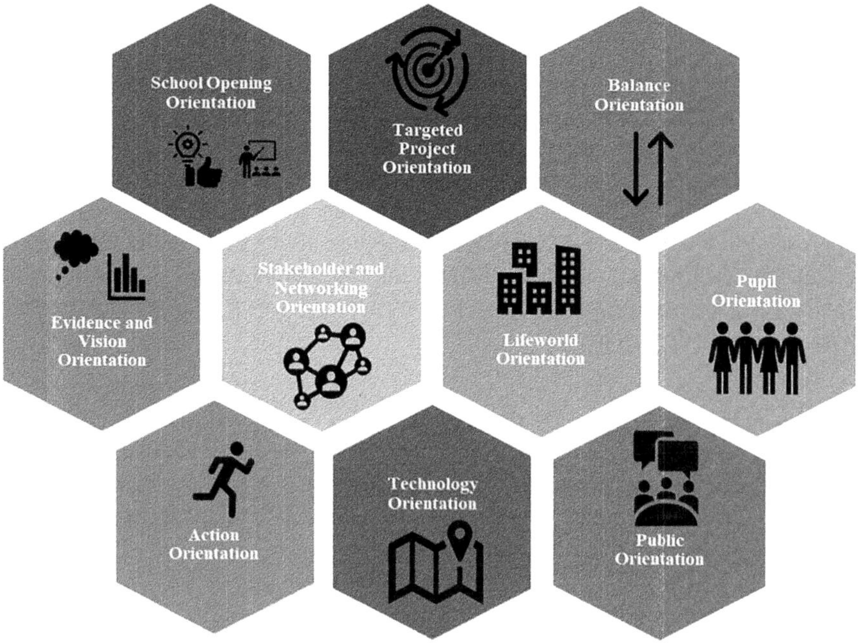

Fig. 13.3 Ten design principles for designing real-world participatory school projects

13.3.1 School Opening Orientation

The initial overarching principle emphasizes integrating schools into genuine participatory structures, thereby positioning them as essential bridges between education and the real world (Mack, 2015). Schools provide a secure environment in which young adults can gain initial, low-barrier participation experiences while developing their participatory competencies. Research indicates that theoretical approaches to teaching democratic action are inadequate; instead, direct application in real-life contexts has a more lasting impact (Pokraka & Gryl, 2018). Moreover, as inclusive institutions that are open to all young adults from a range of social backgrounds due to the compulsory nature of their attendance, schools can engage and empower all young adults, thereby making participatory efforts significantly more inclusive and beneficial.

> In your role as an educator and facilitator between society and the pupils, collaborate with your pupils to develop ideas for participating in local discourses that align with the curriculum (e.g., choice of topic). Establish contact with local politicians to facilitate this engagement.

13.3.2 Targeted Project Orientation

This principle focuses on the design and structure of the participatory environment, drawing on Hart's ladder of participation. The objective is to establish a comprehensive and goal-oriented project framework that encompasses the initial stages of urban planning, such as the identification of problems, and that can be integrated seamlessly into school structures (Bruselius-Jensen & Nielsen, 2021). The project approach circumvents the overburdening of young adults observed in urban planning contexts (Rieker et al., 2016) by creating transparent temporal, thematic, and spatial boundaries and allowing them to contribute their own interests and perspectives (Bruselius-Jensen & Nielsen, 2021).

Create participatory projects that fit within school structures, such as project-oriented learning lessons, working groups, and school project weeks. Focus on a clearly defined geographic topic that pupils can explore within set time limits. Choose topics that not only help them understand the current situation but also encourage them to develop solutions.

13.3.3 Balance Orientation

The objective of this principle is to reinforce a balance between top-down and bottom-up approaches within the context of the project framework. This approach provides young adults with a structured environment (top-down) while allowing them the freedom to explore their own ideas and concerns (bottom-up). Furthermore, it addresses the limited opportunities that young adults often have for meaningful participation in traditional formats and acknowledges that the complexity of planning issues can be overwhelming without adequate support (Rieker et al., 2016). Furthermore, inquiry-based learning, which combines structured guidance (e.g., scientific path of knowledge) with flexibility (e.g., choice of research methods and questions), appears to be an appropriate approach to achieving this balance.

As an advisor and supporter, collaborate with your pupils to develop a project framework based on inquiry-based learning that enables them to explore their own interests and questions that are aligned with their abilities. Structure the project following the geographic–scientific path of knowledge, provide guidance through content and methodological inputs, and include feedback phases to support their progress.

13.3.4 Evidence and Vision Orientation

This principle delineates the methodological design of the project, introducing young adults to a variety of geographical methods that are also employed in planning discourses. These methods should strike a balance between evidence-based argumentation and visionary thinking, as both approaches are considered crucial for developing innovative and realistic solutions. Consequently, the project should include evidence-oriented phases, given that political decision-making often relies on solid data (Hecker et al., 2018). To empower young adults to contribute meaningfully to real-world issues, the citizen science approach can be employed, meaning to incorporate everyday knowledge into scientific knowledge by utilizing scientific research methods (Shirk & Bonney, 2018). However, a purely scientific approach may not fully capture young adults' concerns. To address this issue, visionary utopian thinking is encouraged in alignment with the growing emphasis on solution-focused approaches in geography education (Applis et al., 2022). The future workshop concept (Jungk & Müllet, 1989), which is commonly used in urban planning, can facilitate this by encouraging and guiding young adults to develop visionary ideas in a creative manner.

Develop citizen science projects that emphasize evidence-based research, either by creating personal projects or by participating in existing projects that seek young collaborators. Complement these with phases that encourage visionary thinking. Employ traditional geographical methods, such as mapping and surveys, alongside innovative and creative approaches, such as videography or tracking.

13.3.5 Stakeholder and Networking Orientation

To optimize the sustainable impact of the participation project, it is advisable to integrate a diverse range of stakeholders at all stages of the project's lifecycle. The objective of this principle is to establish a local, multi-perspective educational network that includes scientists, local initiatives, and associations, which could act as valuable links to the political spheres and reinforce young adults' concerns in local decision-making processes. Furthermore, collaboration with different schools is beneficial. Involving a variety of actors contributes diverse knowledge, experiences, and insights to the project, thereby enriching it in a meaningful way (Kenreich, 2018). Simultaneously, this introduces young adults to these institutions, fostering their future engagement.

Engage with local initiatives, NGOs, or scientists who may be relevant to the project. Collaborate with other schools to strengthen the project's impact.

13.3.6 Lifeworld Orientation

This principle is aligned with key ESD principles in geography education, ensuring that the participatory project is relevant to young adults both spatially and thematically. With regard to space and the complexity of urban planning, the project should focus on local scales to enhance accessibility and transparency (Rieckmann, 2021). Regarding the planning theme (e.g., playground design, development, and design of urban mobility concepts), the project should address issues that directly impact young adults' lives (Rieker et al., 2016), extending beyond traditional school topics.

Select topics that directly affect pupils' daily lives by involving them in brainstorming during project planning. Operate on a local scale (e.g., neighborhood) to make pupils' concerns more concrete and foster genuine participation experiences.

13.3.7 Pupil Orientation

This principle underscores the need to engage young adults from a multiplicity of social backgrounds, interests, and perspectives in participation structures to address reported selective engagement (Tisdall, 2021). It is recommended that participation initiatives involve entire classes or multiple grade levels. This approach allows for the integration and consideration of a variety of perspectives and expertise related to the topic. Furthermore, focusing on the young adults' abilities and needs also empowers their self-referential (Rieckmann, 2021) and self-determined learning (Sukuma Arts e.V., n.d.).

Address the needs of your pupils by recognizing them as experts of their own world. Involve a diverse group by opening the project to different grades or classes whenever possible.

13.3.8 Action Orientation

This principle is aligned with ESD teaching principles and is becoming increasingly pertinent in urban planning discourses (Meyer & Haubner, 2022). It motivates young adults to engage actively with spatial issues through practical methods (hands on) that facilitate more profound and sustainable learning experiences (Blum et al., 2021). Consequently, the approach enhances accessibility to intricate planning topics and procedures.

Include phases in which pupils can actively explore urban spaces using action-oriented geographic methods, such as mapping or questioning. Consider incorporating everyday spatial practices (e.g., using bikes).

13.3.9 Technology Orientation

Digital media provides new ways of participation (Allaste & Cairns, 2021), and it is imperative to effectively harness this potential, which will be ensured by this principle. In fact, geography education is particularly well suited to the integration and utilization of geomedia and technologies in participatory projects, given its emphasis on spatial data (Gryl & Jekel, 2012). Furthermore, the utilization of digital geomedia facilitates the comprehension of intricate topics and enables young adults to integrate spatial data with tangible, content-related information (Baker et al., 2018). This is because these technologies provide new data collection and space production methods. Ensure that young adults are competent in the production and utilization of digital geomedia and data in alignment with the tenets of SCE.

Incorporate digital (geo)media, such as global positioning systems (GPS), ArcGIS tools, and open data applications in your project.

13.3.10 Public Orientation

In numerous school or urban planning initiatives involving young adults, their contributions frequently lack sufficient public visibility (Rieker et al., 2016). The objective of this principle is to maintain high visibility and outreach for their work. This can be achieved by disseminating the results of the young adults' work to the public, highlighting their contributions, and thus garnering appreciation (Kenreich, 2018). Potential strategies may include public forums, presentations, and the utilization of social media platforms or websites.

Present the project and its outcomes to the public, engage with politicians and the community, organize presentation events and invite the press, utilize social media channels, and create project websites to enhance visibility.

13.4 The School Project "Verkehrswende Gestalten": Application of the Design Principles to a Project Concerning the Mobility Transition in the Ruhr-Area Metropolis, Essen

Despite the acknowledged necessity for the participation of young adults in urban planning, it has become evident that such endeavors are frequently inadequate, particularly with regard to intricate and regulated matters such as mobility planning (Ohl, 2009).

Numerous cities have been designed with a focus on ease of automobile transportation, which limits independent and safe mobility, particularly for young adults (Ataol et al., 2019), who frequently use bikes for urban mobility and have specific requirements according to the design of streets (Krause & Fuchs, 2021). The imperative for a mobility transition in the context of climate change, with the aim of enhancing quality of life and reducing emissions, underscores the need for the development of improved bike infrastructure. Municipal efforts have not been successful within this transition (Heymann & Eschenfelder, 2019). It is suggested that planners and policymakers fail to consider the needs and requirements of especially young adults in terms of the realization of the mobility transition. The integration of young adults' meaningful perspectives and experiences can yield innovative and suitable solutions for the mobility transition. Furthermore, schools frequently fail to emphasize the requisite skills for effective participation, thereby impeding young adults' engagement in these crucial matters. It was thus decided to develop a project concerning the mobility transition in Essen, a city in the Ruhr Area in Germany.

The main objective of this funded (2023–2025) project (funder: Deutsche Bundesstiftung Umwelt (DBU)) is to engage young adults aged 14–17 years from six schools in Essen (Principle 1) in the redesign and rethinking of bike infrastructure to facilitate the required mobility transition. Furthermore, the initiative strives to foster participatory competencies and self-efficacy in political participation, thereby engaging pupils from diverse social and educational backgrounds (Principle 7). In response to an urgent local issue (Principle 6), the project addresses shortcomings in municipal efforts to develop new mobility concepts. A project-based approach (Principle 2) was devised to address this gap, especially focusing on the incorporation of the marginalized perspectives of young adults into local discourses regarding mobility transition.

The project is structured in three phases (Principle 3; see Fig. 13.4): a kick-off event, a flexible 5-day project unit at each school, and a concluding public event (Principle 10). During the inaugural event, pupils attend a future workshop and create visions for a bike-friendly city (Principle 4). The subsequent project unit can be flexibly integrated into regular geography lessons, project-oriented courses, and even a school project week in accordance with the structures and curriculum of each school. Its objective is to identify and enhance unsafe or unfriendly bike routes based on the pupils' experiences (Principle 4) and data collection. Accordingly, it takes the citizen science approach and follows inquiry-based learning, addressing

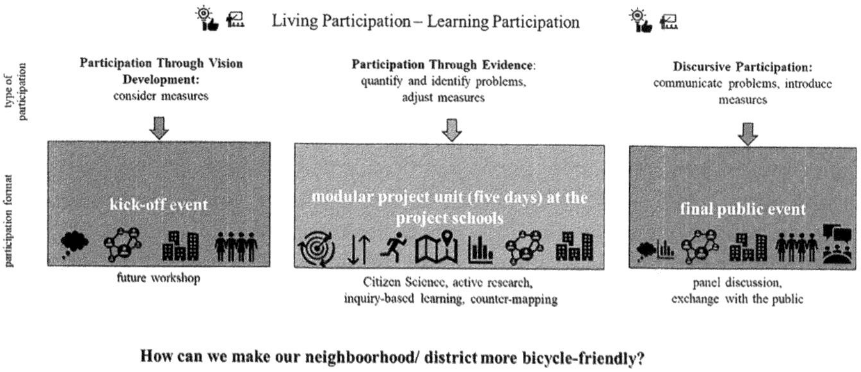

Fig. 13.4 Structure of the developed project

the pupils' questions, such as "What are the conditions of the bike lanes I typically use?" To explore these questions, the senseBox:bike toolkit was developed, customized, and made available to the project by geoinformatics experts at the University of Münster. Prior to collecting spatial and environmental geodata using this device, the pupils also learn to program it (Principle 9). Data collection is performed by attaching the device to the pupils' bikes before they ride through their neighborhood (Principle 8). Additional methods, such as surveys and action camera recordings, are employed (Principle 4). Subsequently, the pupils analyze the datasets (Principle 9) before proposing improvements based on their findings, experiences, and visionary ideas (Principle 4). Then, they present and discuss their findings with local politicians at the final public event (Principle 10).

Throughout the duration of the project, the pupils receive support from various stakeholders (Principle 5), including university scientists who introduce them to the senseBox:bike toolkit, a local planning office, and a local bike NGO. These local stakeholders also provide guidance to the project team by contributing local knowledge and offering pupils insights into their voluntary activities. The project results will be shared with the stakeholders and published on a project website. To enhance visibility and further embed young adults' perspectives, a social media account was created to provide additional insights into pupils' activities and results (Principle 10).

13.5 Living and Learning Participation as a New Approach Taken in Geography Education?!

It should be evident that young adults' participation in intricate societal matters is insufficient and necessitates enhancement, both in the present and especially in the future. To enhance participation, it is essential to create opportunities that facilitate and provide positive participation experiences (living participation) while reducing barriers for young adults to engage in public affairs (learning participation). The creation of such a combined framework in the light of geography education could address this need.

In light of the aforementioned project (Sect. 13.4) which constitutes a pilot project in this field, it is imperative to evaluate which of the presented and practically implemented design principles are of significance and are necessary for the participating pupils and teachers, with the aim of connecting geographical learning with young adults' involvement in real-life urban planning issues. This qualitative evaluation follows the CIPP evaluation model (Stufflebeam & Zhang, 2017) by interviewing participants of the project and compilating interview trends for the adaptation and final preparation of the developed project structures. In line with these findings, educational guidelines and teacher training programs will be developed to implement this specific project or even the derived design principles in different project contexts to strengthen open and life-oriented ESD and transformative learning approaches in geography education.

It is imperative that we cultivate a new generation of spatial citizens, equipped with the ability to collectively and sustainably navigate the complexities and challenges of especially the future world. To this end, it is vital that we prioritize the advancement of learning opportunities based on the principles of living and learning participation within the sphere of geography education.

References

Allaste, A.-A., & Cairns, D. (2021). Digital participation and digital divides in a former socialist country. In M. Bruselius-Jensen, I. Pitti, & E. K. M. Tisdall (Eds.), *Young people's participation: Revisiting youth and inequalities in Europe* (pp. 99–117). Policy Press.

Applis, S., Mehren, R., & Ulrich-Riedhammer, E. M. (2022). Nachhaltigkeit und Ethisches Lernen im Kontext einer lösungsorientierten Didaktik. In D. Mirka, G. Gudat, & J. Laub (Eds.), *Ethik für die Geographiedidaktik: Orientierungen in Forschung und Praxis* (pp. 107–127). transcript Verlag.

Ataol, Ö., Krishnamurthy, S., & van Wesemael, P. (2019). Children's participation in urban planning and design: A systematic review. *Children, Youth and Environments, 29*(2), 27–50.

Baker, T., Curtis, M., & Millsaps, L. (2018). Geotechnologies and the spatial citizen. In E. E. Shin & S. W. Bednarz (Eds.), *Spatial citizenship education: Citizenship through geography* (pp. 117–131). Routledge.

Blum, J., Fritz, M., Taigel, J., Singer-Brodowski, M., Schmitt, M., & Wanner, M. (2021). *Transformatives Lernen durch Engagement: Ein Handbuch für Kooperationsprojekte zwischen Schulen und außerschulischen Akteur*innen im Kontext von Bildung für nachhaltige Entwicklung*. Umweltbundesamt.

Brand, U., & Wissen, M. (2017). Social-ecological transformation. In D. Richardson, N. Castree, M. Goodchild, A. Kobayashi, W. Liu, & R. A. Marston (Eds.), *International encyclopedia of geography: People, the earth, environment, and technology*. Wiley.

Bruselius-Jensen, M., & Nielsen, A. M. W. (2021). The participation project: How projects shape young people's participation. In M. Bruselius-Jensen, I. Pitti, & E. K. M. Tisdall (Eds.), *Young people's participation: Revisiting youth and inequalities in Europe* (pp. 119–135). Policy Press.

Bruselius-Jensen, M., Tisdall, E. K. M., & Pitti, I. (2021). Revisiting young people's participation: An introduction. In M. Bruselius-Jensen, I. Pitti, & E. K. M. Tisdall (Eds.), *Young people's participation: Revisiting youth and inequalities in Europe* (pp. 1–13). Policy Press.

Gryl, I., & Budke, A. (2016). Bildung für nachhaltige Entwicklung - zwischen Utopie und Leerformel? Potentiale für die Politische Bildung im Geographieunterricht. In A. Budke & M. Kuckuck (Eds.), *Politische Bildung im Geographieunterricht* (pp. 57–75). Franz Steiner Verlag.

Gryl, I., & Jekel, T. (2012). Re-centring geoinformation in secondary education: Toward a spatial citizenship approach. *Cartographica: The International Journal for Geographic Information and Geovisualization, 47*(1), 18–28.

Hart, R. A. (1992). *Children's participation: From tokenism to citizenship*. Innocenti Essays.

Hecker, S., Haklay, M., Bowser, A., Makuch, Z., Vogel, J., & Bonn, A. (2018). Innovation in open science, society and policy - setting the agenda for citizen science. In A. Bowser, A. Bonn, J. Vogel, M. Haklay, S. Hecker, & Z. Makuch (Eds.), *Citizen science: Innovation in open science, society and policy* (pp. 1–23). UCL Press.

Heymann, E., & Eschenfelder, C. (2019). *Myths around the transport turnaround: The promise of clean skies*. Deutsche Bank Research. https://www.dbresearch.com/PROD/RPS_DE-PROD/PROD0000000000500325/Myths_around_the_transport_turnaround%3A_The_promise.PDF?undefined&realload=QjlG4cIddVFfgIZDoaz8Q208voDCtSmBVHcIRxU9y8yVzKzTLk3qtcaILHe~2e~1

Hinde, E. R. (2018). Informed citizenry starts in the preschool and elementary grades-and with geography. In E. E. Shin & S. W. Bednarz (Eds.), *Spatial citizenship education: Citizenship through geography* (pp. 132–143). Routledge.

Israel, S., Deakin, J., Franc, R., Markina, A., Murakas, R., & Quandt, M. (2021). Bounded agency and social participation: How socioeconomic situation and experiences influence young people's way of engaging in society. In M. Bruselius-Jensen, I. Pitti, & E. K. M. Tisdall (Eds.), *Young people's participation: Revisiting youth and inequalities in Europe* (pp. 53–76). Policy Press.

Jo, I. (2018). Spatial citizenship in secondary geography curriculum. In E. E. Shin & S. W. Bednarz (Eds.), *Spatial citizenship education: Citizenship through geography* (pp. 145–158). Routledge.

Jungk, R., & Müllet, N. R. (1989). *Zukunftswerkstätten: Mit Phantasie gegen Routine und Resignation*. Heyne.

Kenreich, T. W. (2018). Rediscovering the local: Collaborative, community maps for civic awareness. In E. E. Shin & S. W. Bednarz (Eds.), *Spatial citizenship education: Citizenship through geography* (pp. 72–87). Routledge.

Krause, J., & Fuchs, S. (2021). Belange von Kindern in der Verkehrsplanung. In D. Bär, R. Roth, & F. Csaki (Eds.), *Handbuch kinderfreundliche Kommunen: Kinderrechte kommunal verwirklichen* (pp. 496–504). debus pädagogik.

Mack, W. (2015). Bildung in der Stadt in sozialräumlicher Perspektive. In T. Coelen, A. J. Heinrich, & A. Million (Eds.), *Stadtbaustein Bildung* (pp. 215–221). Springer Fachmedien Wiesbaden GmbH.

Meyer, C., & Haubner, R. (2022). "WIR reden MIT" - Jugendliche partizipieren an der Entwicklung von Lösungsansätzen für kommunale Zielkonflikte der Nachhaltigkeit. In J. Abt, L. Blecken, S. Bock, J. Diringer, & K. Fahrenkrug (Eds.), *Von Beteiligung zur Koproduktion: Wege der Zusammenarbeit von Kommune und Bürgerschaft für eine zukunftsfähige kommunale Entwicklung* (pp. 193–222). Springer Fachmedien Wiesbaden.

Ohl, U. (2009). *Spielraumerweiterung: Institutionelle Rahmenbedingungen und Akteursstrategien in der großstädtischen Stadtteilentwicklung unter Einbezug von Kindern und Jugendlichen*. Doctoral Dissertation, Pädagogische Hochschule Heidelberg. https://opus.ph-heidelberg.de/frontdoor/deliver/index/docId/26/file/Diss_Ohl_Onlineveroeff_15_9_09.pdf

Pokraka, J., & Gryl, I. (2018). Kinder:Karten:Kommunikation - Spatial citizenship zwischen Partizipation und Paternalismus. *KN Journal of Cartography and Geographic Information, 68*(3), 140–146.

Rieckmann, M. (2021). Bildung für nachhaltige Entwicklung: Ziele, didaktische Prinzipien und Methoden. *Merz- Zeitschrift Für Medienpädagogik, 65*(4), 10–17.

Rieker, P., Mörgen, R., Schnitzer, A., & Stroezel, H. (2016). *Partizipation von Kindern und Jugendlichen: Formen, Bedingungen sowie Möglichkeiten der Mitwirkung und Mitbestimmung in der Schweiz. Kinder, Kindheiten, Kindheitsforschung: Band 15*. Springer VS.

Schulze, U., Gryl, I., & Kanwischer, D. (2015). Spatial citizenship education and digital geomedia: Composing competences for teacher education and training. *Journal of Geography in Higher Education, 39*(3), 369–385.

Shin, E. E., & Bednarz, S. W. (2018). Conceptualizing spatial citizenship. In E. E. Shin & S. W. Bednarz (Eds.), *Spatial citizenship education: Citizenship through geography* (pp. 1–9). Routledge.

Shirk, J. L., & Bonney, R. (2018). Scientific impacts and innovations of citizen science. In A. Bowser, A. Bonn, J. Vogel, M. Haklay, S. Hecker, & Z. Makuch (Eds.), *Citizen science: Innovation in open science, society and policy* (pp. 41–51). UCL Press.

Staatsinstitut für Schulqualität und Bildungsforschung München. (n.d.) *Partizipationskompetenz*. Retrieved October 30, 2023, from Lehrplan PLUS. https://www.lehrplanplus.bayern.de/fachprofil/textabsatz/122419

Stufflebeam, D. L., & Zhang, G. (2017). *The CIPP evaluation model: How to evaluate for improvement and accountability*. The Guilford Press.

Sukuma Arts e.V. (n.d.) *BNE- Prinzipien*. Retrieved May 8, 2023, from Meine-BNE.de. https://meine-bne.de/home/expertinnen/prinzipien

Tisdall, E. K. M. (2021). Meaningful, effective and sustainable? Challenges for children and young people's participation. In M. Bruselius-Jensen, I. Pitti, & E. K. M. Tisdall (Eds.), *Young people's participation: Revisiting youth and inequalities in Europe* (pp. 217–234). Policy Press.

UNESCO. (n.d.) *Transformative education: Education for sustainable development and global citizenship education*. Retrieved October 30, 2023, from UNESCO Beirut. https://www.unesco.org/en/fieldoffice/beirut/transformative-education

Lisa Wieczorek has been a research assistant and PhD student at the University of Münster (Germany) since 2023, within the DBU-funded project "Verkehrswende gestalten." Her work focuses on innovative learning formats to enhance youth participation, such as citizen science and inquiry-based learning approaches in geography education, aligned with transformative learning and education for sustainable development. She also investigates STEM education, emphasizing scientific methods and digital geomedia in geography teaching.

Nina Scholten has been based at the University of Hildesheim (Germany) since 2024 and is a professor of geography education. Her research focuses on instructional quality, teacher professionalization, and education for sustainable development. She is dedicated to preparing pre-service teachers to deliver engaging and impactful geography lessons.

Natalie Bienert is currently a trainee teacher at Wilhelm-Hittorf-High School in Münster, Germany, specializing in English and Geography, and maintains close ties to the University of Münster. Her research interests include instructional quality and teacher professionalization in geography education, as well as basic conceptual thinking and learning processes.

Chapter 14
Geography Teaching in Prevention and Protection Against Natural Disasters in Serbia

Ana Milanović Pešić, Marko V. Milošević, and Jovana Brankov

14.1 Introduction

14.1.1 Natural Disasters as a Modern Global Challenge

In recent periods, more people have been affected by natural disasters, which caused significant environmental, social, economic, and cultural consequences (Brooks et al., 2001; Perry, 2018). During the twentieth century, our understanding of natural disasters changed, which resulted in different protection measures against them. Until 1950, the fundamental questions were related to physical processes, while in the second half of the twentieth century, the social aspect was also implemented, and they are no longer viewed only as a natural phenomenon. Since the end of the twentieth and in the twenty-first centuries, special attention has been paid to the consequences of natural disasters and to the possibility of reducing the impact of natural disasters on society, especially in low-income countries (Smith & Petley, 2009).

Based on the Emergency Event Database (EM–DAT database), from 2001 to 2020, 7428 natural disasters occurred, in which 1,209,963 fatalities and 3192.7 billion US$ of economic damage were recorded. In this period, about 364 natural disasters, with about 64,399 fatalities, were registered every year in the world. Economic losses amounted to 200.1 billion US$, mostly from floods (82.7 billion US$, 41.3%) and earthquakes (67.5 billion US$, 33.7%) (CRED, 2023).

A. Milanović Pešić (✉) · M. V. Milošević · J. Brankov
Geographical Institute "Jovan Cvijić" of the Serbian Academy of Sciences and Arts, Belgrade, Serbia
e-mail: a.milanovic@gi.sanu.ac.rs; m.milosevic@gi.sanu.ac.rs; j.brankov@gi.sanu.ac.rs

M. Hardouin, C. Leininger-Frézal (eds.), *Curricular Changes in Geography Education*, International Perspectives on Geographical Education,
https://doi.org/10.1007/978-3-031-96551-7_14

Considering the increasing number of natural disasters with devastating effects, there is a growing interest in disaster-related research and the application of appropriate and novel risk-reduction measures. In this context, the United Nations declared the International Decade for Natural Disaster Reduction (1990–1999) and established the United Nations Office for the Reduction of Risks of Natural Disasters. Three World Conferences (Yokohama, 1994; Kobe, 2005; and Sendai, 2015) on Natural Disaster Reduction emphasized using knowledge and education to improve socio-economic security and resilience, aiming to reduce global disaster mortality, affected populations, and economic damage.

14.1.2 The Role and Importance of Geography in Education About Natural Disasters

In education about natural disasters, the geographical aspect is crucial. Various scholars believe that geography has a leading role in this area because it studies natural disasters and their connections with people, enabling a multidisciplinary approach (Değirmenci & İlter, 2013; Mönter & Otto, 2018). Mitchell et al. (2009, p.172) point out that "geography is a natural academic 'home' to teach about hazards," as "it studies both the natural as well as the social aspect." In addition, geography examines natural hazards and disasters by applying tools such as maps and GIS (Dahlman & Renwick, 2014; Ocak et al., 2023). In many countries, studies of geography education and students' attitudes toward natural disasters have been conducted (Barrow & Haskins, 1993; Selby & Kagawa, 2012).

The analysis of Geography Curricula in many countries showed a lack of content that stimulates proactive action in children during exposure to natural disasters. Komac et al. (2013) concluded that some natural hazards are often inadequately presented in Geography Curricula, considering their territorial representation (e.g., volcanism). In contrast, some countries, such as Turkey, Japan, and Russia, have introduced content related to natural disasters through their Curricula of Geography and other subjects, such as Biology, Physics, and Physical Education (Kartal et al., 2018). Through these Curricula, these countries indicated territorial exposure to certain natural disasters and measures that should be taken preventively or reactively.

Considering that Serbia is exposed to many natural disasters throughout the year, this study examines students' attitudes about natural disasters after the 2014 May Flood, the most devastating natural disaster that hit Serbia in the twenty-first century. To our knowledge, this is the only survey conducted among students after that natural disaster. The obtained results served us to analyze Geography Curricula from that period to determine whether students have functional knowledge about natural disasters.

14.2 Natural Disasters in Serbia

According to the EM–DAT database, Serbia recorded a positive trend in the number of natural disasters in the second half of the twentieth and in the twenty-first centuries. From 2000 to 2023, floods prevailed (60%) among natural disasters, followed by climate-meteorological disasters (37%) (Fig. 14.1).

According to the EM-DAT database and available scientific studies, the biggest natural disasters by fatalities in Serbia from the second half of the twentieth century to the present are the 1981 Flood (70 fatalities) and the 2014 Flood (51 fatalities). In addition, 574 people died in Serbia from extremely high temperatures during May–September 2022. Based on the economic loss, the biggest natural disasters in Serbia are the 2014 Flood (2172 billion US$), the 1990 Drought (1000 billion US$), and the 1965 May Flood (347 billion US$). In Serbia, floods are the most frequent and devastating natural disasters. According to the analyses in 1961–2010, riverine floods in Serbia mainly occur in late spring and early summer (May–June) (Gavrilović et al., 2012). The same trend was observed for the torrential flood occurrence, with the primary peak in May and June and the secondary peak in March (Petrović et al., 2015). From a social aspect, underdeveloped, peripheral municipalities in Serbia have the highest vulnerability to natural disasters (Panić, 2022).

The topics of natural disasters in Serbia and education aimed at risk reduction have been the subject of several studies in recent years (Kovačević-Majkić et al., 2014; Jakovljević et al., 2015). Some studies also included surveys conducted for specific disasters (Panić et al., 2013; Cvetković & Planić, 2022).

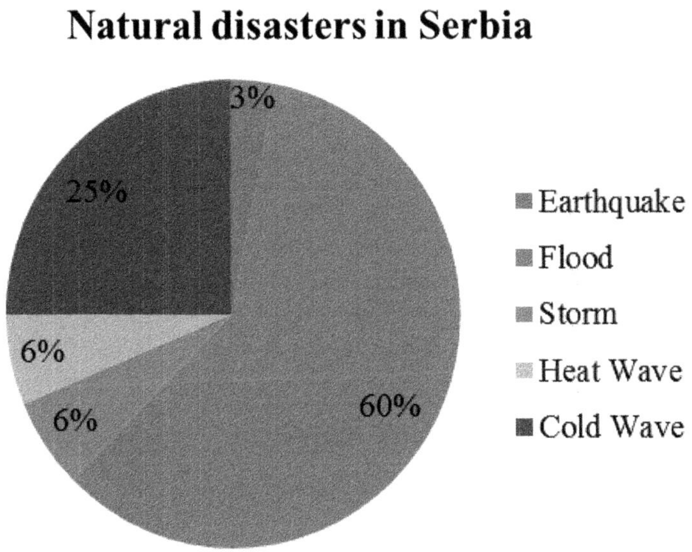

Fig. 14.1 Natural disasters in Serbia in 2000–2023 (source of data: CRED, 2023; created by Milanović Pešić, A.)

14.3 Natural Disasters and Geography Education in Serbia

14.3.1 The Position of Geography in the Education System of Serbia

The education process in Serbia is divided into three primary cycles (Fig. 14.2) and university (Official Gazette RS (OG RS) from 72/2009 to 35/2015). Preschool education is legally mandatory from the sixth to the seventh year. In primary education, classroom teaching is conducted in the lower grades (1st–4th grades) and subject teaching in the higher grades (5th–8th grades) (OG RS from 50/1992 to 72/2009). In secondary education, students choose between gymnasium (acquiring general education) and vocational secondary schools (acquiring specialized knowledge) (OG RS from 50/1992 to 72/2009). Students are introduced to natural objects, processes, and phenomena during the entire educational process. In the lower grades of primary education, their knowledge is at the cognitive level (occurrence of floods, fires, etc.). In the higher grades, natural phenomena are taught at a conceptual level, including their definition, structural elements, and main genetic factors. In secondary education, students acquire broad knowledge about natural phenomena, which enables them to apply knowledge and analyze the process itself.

During primary education in Serbia, the contents related to natural disasters are included in the subjects World Around Us (1st and 2nd grades), Nature and Society (3rd and 4th grades), and Geography (from the 5th to the 8th grades). In secondary education, geography is studied for three years in gymnasiums and one, two, or three years, depending on the vocational schools.

14.3.2 Education About Natural Disaster

In Serbia, there are three forms of education about natural disasters: informal, non-formal, and formal education (Pavićević & Petrović, 2015) (Fig. 14.3).

Forms of non-formal education include seminars, training, courses, and thematic actions. The organizations involved in non-informal education about natural disasters in Serbia are scientific and educational institutions (Faculty of Mining and Geology, Faculty of Geography, Faculty of Physics, Geographical Institute

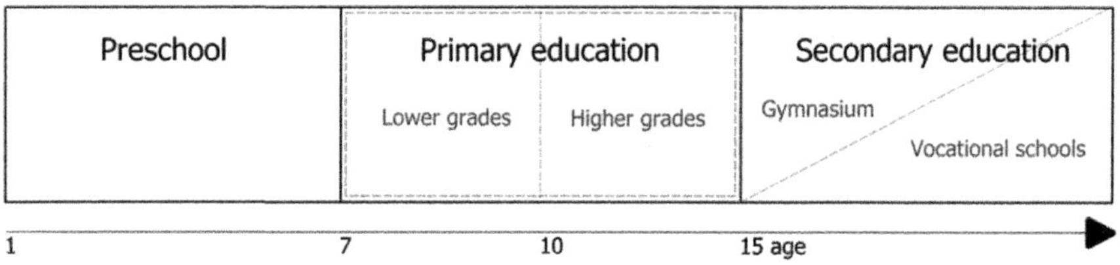

Fig. 14.2 Primary and secondary education in Serbia (created by Milošević, M.)

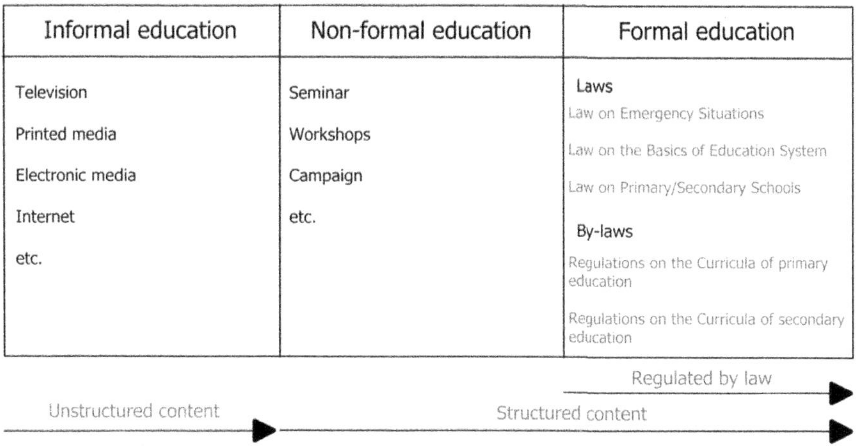

Fig. 14.3 Education forms about natural disasters (created by Milošević, M.)

"Jovan Cvijić" SASA, etc.) and professional institutions (Seismological Institute of Serbia, Republic Hydrometeorological Institute of Serbia, UNICEF, etc.). A good example of non-formal education was the accredited seminars for teachers' education organized by the Geographical Institute "Jovan Cvijić" SASA from 2008 to 2018, focusing on natural disasters, geography teaching, and disaster protection. However, non-formal education includes a small population, mainly based on promoting reactive measures of protection and prevention against natural disasters, which is its main disadvantage.

Formal education allows students to acquire knowledge about natural disasters in general, as well as their spatial and temporal distribution in Serbia. It is regulated by the legal acts of the Ministry of Education and Ministry of Interior Sector for Emergency Management. Until 2018, the legislation, as well as the "National Strategy on Protection and Rescue in Emergency Situations" (Ministry of Interior of the Republic of Serbia, 2011), although they contain general provisions on natural disasters education, was not implemented in Regulations on the School Curricula for subjects World around Us, Nature and Society, and Geography (Milošević et al., 2015).

14.3.3 Geography Curricula (2013–2018)

Based on the desktop analysis of the Curricula (Official Gazette RS – Educational Gazette (OG RS – EG) from 6/2007 to 3/2011; OG RS – EG from 5/2008 to 2/2010; OG RS – EG 10-00-32/97-01; OG RS – EG 1/2006) for the 2013–2018 period, it is concluded that the topic of natural disasters is not included in any of the subjects for primary or secondary education. Geography Curricula included extreme natural phenomena, but only as hazards, not as natural disasters. Earthquakes, volcanism, floods, and precipitation are explicitly mentioned in the Curricula. The only exception is the Geography Curriculum for the 2nd grade of Secondary School of Tourism (OG RS – EG 10/2012, 1/2013), in which two topics related to natural disasters are defined—Water management problems in Serbia and Natural disasters.

After reviewing the contents of the textbook, it was noticed that the most extreme natural disasters in the world were presented, without examples of natural disasters in Serbia. Therefore, the mentioned contents do not have a stimulating role in strengthening the awareness of natural disasters. This conclusion is also confirmed by research after the 2010 Earthquake in Kraljevo when it was determined that over 70% of children did not know an earthquake could happen in their area (Panić et al., 2013). All of the above prompted us to survey the students' perceptions of natural disasters and education after the May 2014 Flood, the biggest recorded natural disaster in Serbia in the twenty-first century.

14.3.4 Education for Protection Against Natural Disasters: Analysis of Survey Research

Survey research was conducted in June 2014 in several municipalities of Šumadija District to examine the understanding, knowledge, and thinking about natural disasters among school children. It was conducted on a voluntary basis and anonymously. The participants and parents were informed about the purpose of the research and that all results would be used exclusively for scientific research. After obtaining the consent of all respondents, the research was conducted in accordance with the current legislation of the Republic of Serbia.

Šumadija District is located in Central Serbia and covers 8569 km^2 of area, with 22 municipalities located entirely in this territory. The May 2014 Flood, which hit 39 municipalities in Central Serbia, was also extended to 12 municipalities of the Šumadija District (Milanović Pešić, 2015). During this flood, the absolute maximum water levels in the instrumental period were reached at many hydrological stations. After the floods, landslides were also activated in several municipalities (Petrović et al., 2016; Milošević et al., 2021). Significant floods in this area were also recorded in 1965, 1969, 1981, and 2011, and one of the biggest occurred in July 1999, when eight people died, several hundred facilities and buildings and 30 bridges were damaged (Gavrilović et al., 2012).

The respondents were students of higher grades (7th–8th grades) of primary schools and students of secondary schools (1st–4th grades). The research included 649 respondents (434 primary school students and 215 secondary school students) from 10 primary and 10 secondary schools. Among them, 229 girls (52.8%) and 205 boys (47.2%) from primary school and 138 girls (64.2%) and 77 boys (35.8%) from secondary school participated in the survey. The data were collected by using a self-administered questionnaire distributed to students during the class, along with instructions for filling it out.

The survey data were analyzed with SPSS software, using Chi-square and Cramer's V coefficient. These methods are often used to examine respondents'

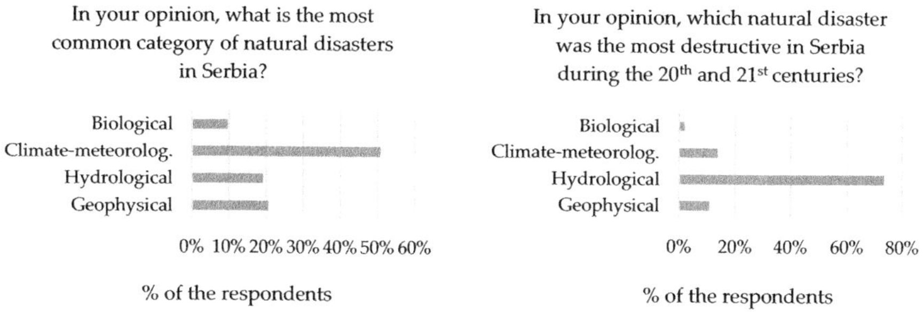

Fig. 14.4 Students' attitudes toward the most common (left) and most destructive (right) natural disaster in Serbia (created by Brankov, J.)

perceptions and attitudes on various topics, including water quality and pollution issues (Milanović Pešić et al., 2020; Brankov et al., 2021) and environmental protection (Alkan et al., 2009; Jojić Glavonjić et al., 2019; Brankov et al., 2022).

The part of the questionnaire analyzed in this chapter consisted of seven questions (Table 14.1). When asked about the most common category of disasters in Serbia, 50.8% of respondents opted for climate-meteorological disasters, 20.5% for hydrological, 19.1% for geophysical and only 9.6% for biological disasters. When asked to highlight the most devastating category of disaster, most students chose hydrological disaster (73.5%), compared to the students who singled out climate-meteorological (13.7%), geophysical (10.8%), and biological (2%) disasters (Fig. 14.4).

According to the students, climate-meteorological disasters occur most often in their municipality (56.3%), followed by hydrological (19.1%), geophysical (14.5%), and biological (10.1%) disasters. Regarding the most destructive category of disasters experienced by respondents, 45% believe that it was a hydrological disaster, while 38.6% opted for a climate-meteorological disaster. A significantly smaller percentage of respondents singled out some other type of disaster.

More than half of the respondents (55.7%) believe that natural factors caused the disaster they experienced, 31.3% think that natural and anthropogenic factors were the causes, and 13% opted only for anthropogenic factors.

When asked what would contribute to reducing the consequences of natural disasters in their municipality, 40.1% of students highlighted better information about what to do and how to behave during the disaster, 25.2% singled out preventive measures before disasters, 25% opted for greater material investments in protection systems, and 9.7% singled out better work of civil protection. Most respondents emphasized the need for knowledge and training in disaster behavior (34.6%) and after it (55.7%) (Fig. 14.5).

The Chi-square test was applied to determine whether there are statistically significant differences in the distribution of students' attitudes by type of school for all the questions mentioned above. Cramer's V coefficient was used to determine the strength of the link between category variables (Table 14.1).

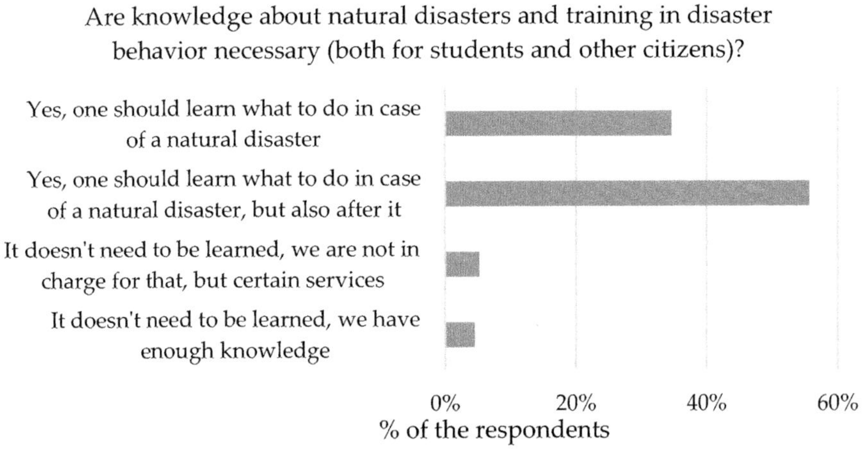

Fig. 14.5 Primary and secondary school students' attitudes toward the need for knowledge and training in disaster behavior (created by Brankov, J.)

Table 14.1 Differences in attitudes of the students about natural disasters (created by Brankov, J.)

Students' attitudes		χ^2 (*p-value*)	Cramer's V coeffic.
The most frequent category of natural disasters in Serbia	Type of school	**0.032**	0.122
The most devastating category of natural disaster in Serbia in the twentieth and the twenty-first centuries	Type of school	0.449	0.067
The most frequent category of natural disasters in your municipality	Type of school	0.190	0.127
The most devastating category of natural disaster you experienced in your municipality	Type of school	**0.001**	0.167
Factors that led to the occurrence of the most devastating disaster in your municipality	Type of school	0.212	0.070
Reducing the effects of natural disasters in your municipality	Type of school	0.199	0.094
The need for knowledge about natural disasters and training in disaster behavior	Type of school	0.587	0.055

Of all the analyzed variables, a link has been established between the respondents' type of school and their attitude about the most frequent category of natural disasters in Serbia. Although both primary and secondary school students singled out climate-meteorological disasters as the most common in Serbia, a significant percentage of primary school students (23.2%) chose hydrological disasters, in contrast to only 14.5% of secondary school students. A higher percentage of secondary school students opted for geophysical disasters (23.7%) compared to 17% of primary school students. In previous case, Cramer's coefficient testifies to the low correlation strength.

Additionally, a correlation is found between the respondents' type of school and their attitude about the most devastating category of natural disaster they experienced in their municipality. Most primary school students (49%) stated that the most destructive disaster they experienced was a hydrological disaster, in contrast to high school students (49.5%) who mostly opted for climate-meteorological disasters.

In general, survey results show that students need more information and functional knowledge about natural disaster protection, and are interested in learning more to enhance their awareness.

Survey research on students' attitudes toward natural disasters was conducted in various countries to assess their knowledge and the topic's representation in Curricula. After the 2008 Sichuan Earthquake in China, a conducted survey among students and teachers in four cities indicated that education about natural disasters should be improved to raise students' awareness and ability to react (Zhu & Zhang, 2017). During several education projects about natural disasters, Piangiamore et al. (2015) have involved more than 20,000 students and teachers in Italy who live in areas prone to natural hazards. Among the activities was a survey filled out by children in order to test a set of learning modules. In Greece, a survey revealed inadequate knowledge among teachers and students of responsibilities and necessities during disasters (Mereli et al., 2022). Based on the findings mentioned above, it can be concluded that the results of our survey align with research in other countries.

14.3.5 Geography Curricula (From 2018)

Based on the desk analysis of Geography Curricula before 2018, it was concluded that students learned about natural disasters only at a conceptual level, without an indication that these phenomena occurred in Serbia and without defining the most threatened areas. There were also no instructions on adequate measures in case of exposure to any natural disaster. All these shortcomings were eliminated in the new Geography Curricula from 2018, and the learning outcomes were changed. As an example, a comparison of the learning outcomes about volcanoes and earthquakes according to the old and new Geography Curriculum for the 5th Grade of Primary Education is given in Table 14.2.

The new Geography Curriculum should enable learning about natural disasters in three aspects: cognitive, affective, and psychomotor. Students should understand disaster processes, dynamics, and distribution (cognitive), act rationally during disasters to reduce panic (affective), and develop practical and digital skills to minimize disaster impacts (psychomotor). Using tools like the Disaster Risk Register made in 2022 by the Ministry of Interior of the Republic of Serbia, Emergency Management Sector, and the Republic Geodetic Authority, they can recognize disaster characteristics and early warnings.

Table 14.2 Topic volcanoes and earthquakes within the Geography Curriculum for the 5th grade of primary education (created by Milošević, M.)

Topic: Volcanoes and earthquakes	
Before 2018[a]	After 2018[b]
1. Student distinguishes and explains the concept of volcanism, earthquake, and volcanic eruption	1. Using maps and drawings, the student describes the ways and consequences of the movement of lithospheric plates (volcanism, earthquakes, folding, and faulting) 2. Student differentiates between epicenter and hypocenter and lists risk zones in the world and in Serbia 3. Students list the procedures to be taken during an earthquake

[a]OG RS—EG from 6/ 2007 to 3/2011
[b]OG RS—EG from 15/ 2018–77 to 14/2023–384

14.4 Conclusion

The conducted survey about the respondents' knowledge of the most common and destructive natural disasters in Serbia revealed that students lack specific knowledge about disasters at the national and local levels, and their responses showed uncertainty. Despite this, they expressed interest in further education through geography classes and extracurricular activities. Statistical analysis of attitudes indicated that school type influences students' opinions.

In order to increase the resilience of the society, especially among students, Serbia established formal education on natural disasters in 2017–18. The Ministry of Education of the Republic of Serbia enacted the Regulation on the Plan for 5th and 6th Grade of Primary Education and the Curriculum for 5th and 6th Grade of Primary Education (OG RS – EG from 15/ 2018-77 to 14/2023-384), which explicitly defines the outcomes related to this thematic field and establishes formal education about natural disasters in Serbia. In addition, training and workshops organized through additional school activities can help educate children to become more aware of the dangers of natural disasters and better informed about these phenomena. In our opinion, besides familiarization with natural hazards on conceptual level, formal education about natural disasters should also indicate: 1. How to react in case of exposure to a certain natural disaster; 2. What are possible natural disasters in the area where students live and what are the periods when they can occur (this does not apply to earthquakes); 3. What are the relevant sources of information for specific natural hazards, whether it is their spatial representation (e.g., Landslide Cadastre) or the event announcement (MeteoAlarm—Storm announcement).

Although much has been done, many outstanding issues need to be explored to gain effective policy and practice. Stronger collaboration among academics, agencies, and different types of community networks is needed to build up an evidence base on the most effective adaptation and mitigation measures through education.

Acknowledgments This research was supported by the Ministry of Education, Science and Technological Development of the Republic of Serbia [Project no. III 47007] and Ministry of Science, Technological Development and Innovation of the Republic of Serbia (Contract No. 451-03-136/2025-03/200172).

The authors express their gratitude to the geography teachers (Ana Bogdanović, Boris Zdravković, Dobrila Lukić, Javorka Jakovljević, Nevenka Bulajić, Olgica Tomić Gaković, Srećko Antunović, Violeta Načković, and Željko Poštić) and all students who participated in the survey.

References

Alkan, H., Korkmaz, M., & Tolunay, A. (2009). Assessment of primary factors causing positive or negative local perceptions on protected areas. *Journal of Environmental Engineering and Landscape Management, 17*(1), 20–27. https://doi.org/10.3846/1648-6897.2009.17.20-27

Barrow, L. & Haskins, S. (1993). Earthquakes haven't shaken college students' cognitive structure. In *Proceedings of the Third International Seminar on Misconceptions and Educational Strategies in Science and Mathematics*. Ithaca, New York, 1st–4th August 1993, Misconceptions Trust, 1–13.

Brankov, J., Milanović Pešić, A., Milijašević Joksimović, D., Radovanović, M. M., & Petrović, M. D. (2021). Water quality estimation and population's attitudes: A multi-disciplinary perspective of environmental implications in Tara National Park (Serbia). *Sustainability, 13*(1), 241. https://doi.org/10.3390/su13010241

Brankov, J., Micić, J., Ćalić, J., Kovačević-Majkić, J., Milanović, R., & Telbisz, T. (2022). Stakeholders' attitudes toward protected areas: The case of Tara National Park (Serbia). *Land, 11*(4), 468. https://doi.org/10.3390/land11040468

Brooks, G. R., Evans, S. G. & Clague, J. (2001). Floods. *Geological Survey of Canada Bulletin 548—A Synthesis of Geological Hazards in Canada*, 101–143.

Centre for Research on the Epidemiology of Disasters (CRED). (2023). Emergency events database. Brussels. Accessed October 8, 2023, from https://www.emdat.be/emdat_db/

Cvetković, V., & Planić, J. (2022). Earthquake risk perception in Belgrade: Implications for disaster risk management. *International Journal of Disaster Risk Management, 4*(1), 69–88. https://doi.org/10.18485/ijdrm.2022.4.1.5

Dahlman, C. T., & Renwick, W. H. (2014). *Introduction to geography: People, places & environment* (6th ed.). Pearson Education Limited.

Değirmenci, Y., & İlter, I. (2013). The natural disasters in the geography teaching curriculum. *Marmara Coğrafya Dergisi/Marmara Geographical Review, 28*, 236–303.

Gavrilović, L., Milanović Pešić, A., & Urošev, M. (2012). A hydrological analysis of the greatest floods in the Serbia in the 1960-2010 period. *Carpathian Journal of Earth and Environmental Sciences, 7*(4), 107–116.

Jakovljević, V., Cvetković, V., & Gačić, J. (2015). In B. Banović (Ed.), *Natural disaster and education*. University of Belgrade/Faculty of Security Studies. (in Serbian).

Jojić Glavonjić, T., Doljak, D., Brankov, J., & Filipović, M. (2019). Residents' perception toward protected areas – Landscape of exceptional features Vlasina (Serbia). *Carpathian Journal of Earth and Environmental Sciences, 14*(1), 5–17. https://doi.org/10.26471/cjees/2019/014/0532

Kartal, V., Orlović Lovren, V., Milošević, M. V., Kovačević-Majkić, J., Ćalić, J., Milanović Pešić, A., Belij, S., Štrbac, D., Panić, M. & Miljanović, D. (2018). *How to protect ourselves against natural disasters – Handbook for primary school teachers]*. : The Primary School Teachers' Association of the Republic of Serbia, UNICEF, Ministry of Education, Science and Technological Development of the Republic of Serbia. ISBN: 978-86-82469-11-7 (in Serbian).

Komac, B., Zorn, M., & Ciglič, R. (2013). European education on natural disasters – A textbook study. *Natural Hazards Earth System Science Discussions, 1*, 2255–2279. https://doi.org/10.5194/nhessd-1-2255-2013

Kovačević-Majkić, J., Milošević, M. V., Panić, M., Miljanović, D., & Ćalić, J. (2014). Risk education in Serbia. *Acta Geographica Slovenica, 54*(1), 163–178. https://doi.org/10.3986/AGS54305

Mereli, A., Evelpidou, N., Antonarakou, A., Drinia, H., Mereli, M., & Tzouxanioti, M. (2022). Investigation of the beliefs and assessment of the security feeling in primary education in Greece in relationship to rapid onset natural disasters. *International Journal of Educational Research, 7*(2), 56–70. https://doi.org/10.24331/ijere.1028563

Milanović Pešić, A (2015). *Geographical aspects of natural disaster in Šumadija*. PhD Dissertation, University of Belgrade–Faculty of Geography (unpublished, in Serbian).

Milanović Pešić, A., Brankov, J., & Milijašević Joksimović, D. (2020). Water quality assessment and populations' perceptions in the National park Djerdap (Serbia): Key factors affecting the environment. *Environment, Development and Sustainability, 22*, 2365–2383. https://doi.org/10.1007/s10668-018-0295-8

Milošević, M. V., Štrbac, D., Ćalić, J., Kovačević-Majkić, J., Panić, M., & Miljanović, D. (2015). Education in the system of prevention and protection against natural disasters. *Izgradnja, 69*(11–12), 547–556. (in Serbian).

Milošević, M. V., Štrbac, D., Ćalić, J., & Radovanović, M. (2021). Detection of earthflow dynamics using medium-resolution digital terrain models: Diachronic perspective of the Jovac earthflow, Southern Serbia. *Acta Geographica Slovenica, 61*(2), 187–206. https://doi.org/10.3986/AGS.9818

Ministry of Interior of the Republic of Serbia. (2011). National strategy on protection and rescue in emergency situations. Accessed October 30, 2023, from https://www.fao.org/faolex/results/details/en/c/LEX-FAOC202287/ (in Serbian).

Mitchell, J. T., Borden, K. A., & Schmidtlein, M. C. (2009). Teaching hazards geography and geographic information systems: A middle school level experience. *International Research in Geographical and Environmental Education, 17*(2), 170–188. https://doi.org/10.1080/10382040802148679

Mönter, L., & Otto, K.-H. (2018). The concept of disasters in geography education. *Journal of Geography in Higher Education, 42*(2), 205–219. https://doi.org/10.1080/03098265.2017.1339266

Ocak, F., Döker, M. F., & Ünsal, Ö. (2023). Use of web-based GIS applications in geography teaching—The implications from Türkiye. *Journal of the Geographical Institute "Jovan Cvijić" SASA, 73*(1), 65–77. https://doi.org/10.2298/IJGI2301065O

Official Gazette RS. (50/1992, 53/1993, 67/1993, 48/1994, 24/1996, 23/2002, 25/2002, 62/2003, 64/2003, 101/2005, 72/2009). The law on secondary schools. https://www.paragraf.rs/propisi/zakon_o_srednjem_obrazovanju_i_vaspitanju.html (in Serbian).

Official Gazette RS. (50/1992, 53/1993, 67/1993, 48/1994, 24/1996, 23/2002, 25/2002, 62/2003, 64/2003, 101/2005, 72/2009). Law on secondary schools. https://www.paragraf.rs/propisi/zakon_o_srednjem_obrazovanju_i_vaspitanju.html (in Serbian).

Official Gazette RS. (72/2009, 52/2011, 55/2013, 35/2015). The law on the basics of education system. https://www.paragraf.rs/propisi_download/zakon_o_osnovama_sistema_obrazovanja_i_vaspitanja.pdf (in Serbian).

Official Gazette RS – Educational Gazette. (1/2006). Regulation on the curricula for 2nd, 3rd and 4th grade of gymnasium (in Serbian).

Official Gazette RS – Educational Gazette. (6/2007, 2/2010, 7/2010, 3/2011). Regulation on the plan for the 2nd cycle of primary education and the curriculum for 5th grade of primary education (in Serbian).

Official Gazette RS – Educational Gazette. (5/2008, 6/2009, 2/2010). Regulations on the curricula for 6th, 7th and 8th grade of primary education (in Serbian).

Official Gazette RS – Educational Gazette. (10-00-32/97-01). Regulation on the plan for the gymnasium education and the curriculum for 1st grade of gymnasium (in Serbian).

Official Gazette RS– Educational Gazette. (10/2012, 1/2013). Regulation on the curriculum for 2nd grade of secondary school of tourism (in Serbian).

Official Gazette RS– Educational Gazette. (15/2018-77, 18/2018-1, 3/2019-83, 3/2020-3, 6/2020-94, 17/2021-1, 16/2022-1, 13/2023-458, 14/2023-384). Regulation on the plan for the 5th and 6th grade of primary education and the curriculum for 5th and 6th grade of primary education (in Serbian).

Panić, M. (2022). *Social vulnerability to natural hazards in Serbia*. In A. Milanović Pešić (Ed.). Special Issues 98, Geographical Institute "Jovan Cvijić" SASA (in Serbian).

Panić, M., Kovačević-Majkić, J., Miljanović, D., & Miletić, R. (2013). Importance of natural disaster education – Case study of the earthquake near the City of Kraljevo. *Journal of the Geographical Institute "Jovan Cvijić" SASA, 63*(1), 75–88. https://doi.org/10.2298/IJGI121121001P

Pavićević, M., & Petrović, D. (2015). Differences between formal, non-formal and informal education. *Zbornik radova Učiteljskog fakulteta, 9*, 103–113. https://doi.org/10.5937/zrufpl1509103P. (in Serbian).

Perry, R. W. (2018). Defining disaster: An evolving concept. In H. Rodríguez, W. Donner, & J. Trainor (Eds.), *Handbook of disaster research. Handbooks of sociology and social research*. Springer. ISBN 978-3-319-63254-4.. https://doi.org/10.1007/978-3-319-63254-4_1

Petrović, A., Dragićević, S., Radić, B., & Milanović Pešić, A. (2015). Historical torrential flood events in the Kolubara river basin. *Natural Hazards, 79*, 537–547. https://doi.org/10.1007/s11069-015-1860-1

Petrović, A. M., Kovačević-Majkić, J., & Milošević, M. V. (2016). Application of run-off model as a contribution to the torrential flood risk management in Topčiderska Reka watershed, Serbia. *Natural Hazards, 82*, 1743–1753. https://doi.org/10.1007/s11069-016-2269-1

Piangiamore, G. L., Musacchio, G., & Pino, N. A. (2015). Natural hazards revealed to children: The other side of prevention. In G. di Capua & S. Peppoloni (Eds.), *Geoethics: The role and responsibility of geoscientists* (pp. 171–181). The Geological Society of London. ISBN: 9781862397262. https://doi.org/10.1144/SP419.12

Selby, D., & Kagawa, F. (2012). *Disaster risk reduction in school curricula: Case studies from thirty countries*. UNESCO, United Nations Educational, Scientific and Cultural Organization, UNICEF, United Nations Children's Fund. Accessed October 25, 2023, from https://resourcecentre.savethechildren.net/document/disaster-risk-reduction-school-curricula-case-studies-thirty-countries/

Smith, K., & Petley, D. N. (2009). *Environmental hazards: Assessing risk and reducing disaster* (5th ed.). Routledge.

Zhu, T. T., & Zhang, Y. J. (2017). An investigation of disaster education in elementary and secondary schools: Evidence from China. *Natural Hazards, 89*, 1009–1029. https://doi.org/10.1007/s11069-017-3004-2

Dr Ana Milanović Pešić is a Senior Research Associate at the Geographical Institute "Jovan Cvijić" SASA and is currently Deputy Director of the Institute. Her research area is focused on physical geography, with a special emphasis on hydrology. She has worked on numerous projects and as a lecturer at professional training programs for geography teachers. She was guest editor of international journals and is editor of the Journal of the Geographical Institute Jovan Cvijić SASA.

Dr Marko V. Milošević was born in 1978 in Jagodina (Serbia). He got PhD of Geographical Sciences (2021) at Faculty of Geography, University of Belgrade. Major field of study was physical geography. Research focus defined as: geomorphology, natural disasters and geographical educational policies. Dr Milošević has been employed since 2003 at Geographical Institute "Jovan Cvijić" of the Serbian Academy of Sciences and Arts at the Department of Physical Geography.

Dr Jovana Brankov is a Senior Research Associate at the Geographical Institute "Jovan Cvijić" SASA. She was engaged as a Senior Researcher at the South Ural State University, Institute of Sports, Tourism and Service, in Chelyabinsk (Russia) (2018-2022). Her scientific direction is focused on broader aspects of social geography, with a special emphasis on sustainable development, sustainable tourism, and community-based research. She participates in numerous projects dealing with social geography and sustainable development issues.

GPSR Compliance

The European Union's (EU) General Product Safety Regulation (GPSR) is a set of rules that requires consumer products to be safe and our obligations to ensure this.

If you have any concerns about our products, you can contact us on ProductSafety@springernature.com

In case Publisher is established outside the EU, the EU authorized representative is:

Springer Nature Customer Service Center GmbH
Europaplatz 3
69115 Heidelberg, Germany

Batch number: 10755292

Printed by Printforce, the Netherlands